FOREWORD

Carnegie Mellon's STUDIO for Creative Inquiry supports atypical, interdisciplinary, and interinstitutional research at the intersection of the arts, science, and technology. In parallel, the Miller Gallery at Carnegie Mellon University supports experimentation that expands the notions of art and culture, providing a forum for engaged conversations about creativity and innovation. Together our units work to develop and present new research in the arts. This publication represents the capstone to a new curatorial residency program developed jointly by the Miller Gallery and the STUDIO, with support from the Andy Warhol Foundation for the Visual Arts.

We are proud to present this timely reflection on the current and historic intersections of art, science, and technology. This book is the type of adventurous and interdisciplinary investigation that we seek to foster. We have been delighted to host Luke Bulman, Régine Debatty, Claire Evans, Pablo Garcia, and Jessica Young, along with principal investigator and curatorial fellow Andrea Grover, who deftly guided this project to completion in a mere seven days.

—Astria Suparak
Director, Miller Gallery at Carnegie Mellon University

—Golan Levin
Director, Carnegie Mellon STUDIO for Creative Inquiry

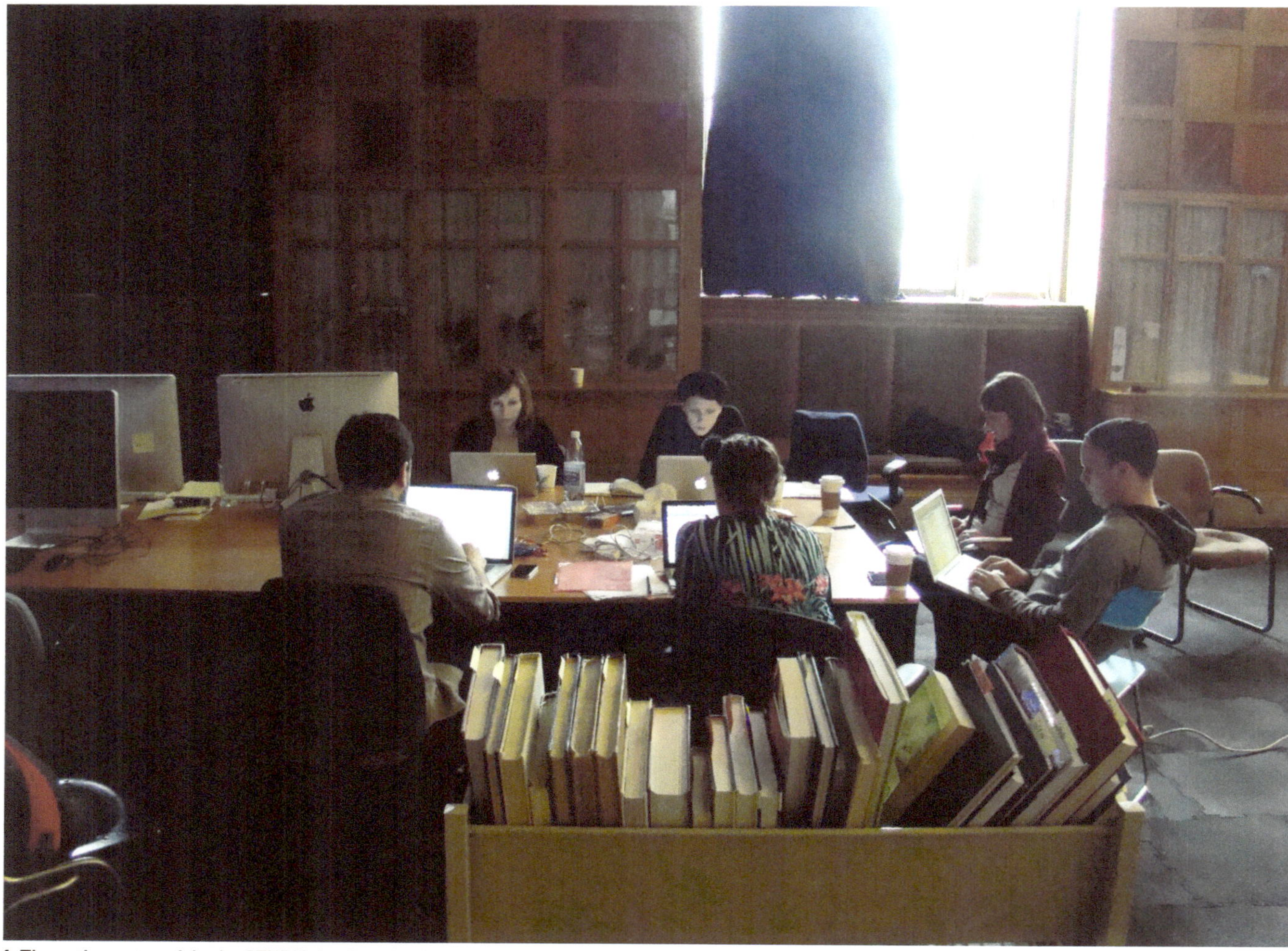

1. The authors at work in the STUDIO for Creative Inquiry, February 18, 2011. Photo: Jonathan Minard

EXTREME WRITING

It's 4:01 p.m. on February 18, 2011. There are ten of us in the STUDIO for Creative Inquiry—a former library now emptied of its books and reconfigured for computers and projection—within the College of Fine Arts at Carnegie Mellon University. The tables in the STUDIO are cluttered with empty food plates, coffee cups, notepads, essays, and publications from the Hunt Library next door. The windows are open and it's 61 degrees Fahrenheit, a rare occurrence for February in Pittsburgh. The room is quiet with the exception of the occasional murmur of conversation and the sound of keyboards ticking. We are five days into a seven-day challenge: to collectively author and design a book on the subject of contemporary artists working at the intersection of art, science, and technology. Despite the appearance of working independently, we are all networked, reading and writing toward the same purpose.

The publication you hold in your hands was created this way over the course of one week (February 14–20, 2011) by four writers and two graphic designers, with the assistance of two readers and eleven work-study students. It was a "book sprint." Derived from "code sprinting," a method for working on an open source project by getting software developers into a single room for a period of intensive work, the term *book sprint* describes the quick, collective writing of a topical book. The process has a long and interdisciplinary lineage: we see the same idea in think tanks, collective intelligence frameworks, telepathy, and the notions of cyborgs and the "metabrain." No one section of this book has a sole author, and the writing process occurred in a nonlinear, simultaneous, and synergistic fashion in the collective workspace at the STUDIO.

The concept of a book sprint isn't ours; we're indebted to book sprint astronauts FLOSS Manuals[1] and the participants in *Collaborative Futures* at transmediale 2010 and 2011[2]–the first people with the inspiration to translate the "sprinting" method to something other than code writing or technical text. This process wouldn't have been possible a decade ago, either: the Internet has provided us with instant access to reference materials, a means to write simultaneously on one text, and the bandwidth to incorporate images and visual design in real time.

We launched *our* book sprint in order to produce a snapshot of this particular moment—and because we wanted to do it with immediacy, without distraction. The topic of this publication is the most recent manifestation of artists working in art, science, and technology, which we broadly define as work that adopts processes of the natural or physical sciences, "does strange things with electricity" (to borrow a phrase from Dorkbot[3]), breaks from traditional models of art/science pairings, and was created within the last five years. We realize that art, science, and technology intersections have a tradition with much deeper roots than we have space to detail here (and that such histories have been given attention elsewhere), so we've provided in a timeline a brief subjective history of innovations, movements, and cultural events that have contributed to this tradition and led us to this moment. To be clear: this book is an effort to understand this very moment in art, science, and technology affinities, and the ways Internet culture and networked communication have shaped the practice.[4]

—Andrea Grover
Project Lead, Warhol Curatorial Fellow at the STUDIO for Creative Inquiry and the Miller Gallery at Carnegie Mellon University

1. FLOSS Manuals is a non-profit online community whose aim is to produce quality free documentation for free software.
2. transmediale is an annual festival for art and digital culture held in Berlin.
3. Dorkbot is a group of affiliated organizations worldwide that sponsors grassroots meetings of artists, engineers, designers, scientists, inventors, and anyone else working under the very broad umbrella of electronic art. The Dorkbot motto is "people doing strange things with electricity."
4. An even larger question, to be considered in another forum, is how Internet culture and networked communication is shaping culture and politics at this moment. During the week we were creating this document, newspapers were placed daily at the threshold of our hotel rooms. Photos of protests from around the world were front page news. The two-week-old "Egyptian Revolution of 2011" had set in motion a worldwide movement by virtue of its visibility and ability to communicate its message instantly and globally.

CONTENTS

INTRODUCTION

"The artist is a positive force in perceiving how technology can be translated to new environments to serve needs and provide variety and enrichment of life. He may be the only one who can transcend cultural bias and deal with the individuals of a culture on their own terms."
—Billy Klüver, engineer and co-founder of Experiments in Art and Technology

The late 1960s is the period most commonly associated with the origins of interdisciplinary collaborations as we know them today. The world had mixed emotions about technology: NASA had placed the first man on the Moon, vaulting astronauts and engineers to rock-star status, while the Vietnam War had advanced the war machine—live on color television. In the spirit of these times, the idea began to surface for artists to intervene and redirect the new technologies.

Between 1966 and 1971, artist collaborations with engineers and scientists reached a fever pitch, embodied in efforts like the Art and Technology (A&T) Program at LACMA,[1] Experiments in Art and Technology (E.A.T.),[2] and the Artist Placement Group.[3] Each of these organizations aimed to pair artists with, or place them within, scientific or industrial environments with the intention of providing them with access to state-of-the-art technologies, the knowledge assets of scientists and engineers, manufacturing processes, and the experience of being embedded in corporate culture.

There were many contributing factors to this transformative moment: the countercultural leanings of the 1960s, a growing interest in system theories[4] (theories inspired by behaviorism and cybernetics), and a desire for artists to intervene in the industrial sector, specifically around technologies associated with warfare. The art historical origins of this moment lie in Russian Constructivism, Futurism, Bauhaus, Situationism, and Fluxus—all movements that sought to more fully integrate art into the social sphere. Physicist and novelist C.P. Snow's now canonized 1959 Rede lecture, "The Two Cultures,"[5] was yet another catalyst of the time. It was Snow's provocation that the breakdown in communication between the sciences and the humanities should be remedied or it would remain a major hindrance to solving the world's problems.

Snow argued that if the so-called two cultures (science and the humanities) couldn't manage to find a way to communicate—or at least overcome their pretensions long enough to respect one another—then the great findings of science and the great works of art would never get the discourse and celebration they deserve. Without a shared language, the frameworks that intellectuals were building on either side of the chasm would only serve to perpetuate the ideology of their own disciplines without adding to the whole. Pamela Lee, in *Chronophobia: On Time in the Art of the 1960s*, contends that "Snow's position was critical in articulating the historical confluence of arts and sciences from the sixties forward: the lecture anticipated, in numerous ways, what would later be described as the phenomenon of interdisciplinarity within academia."[6]

Nostalgia for the hallmark interdisciplinary efforts of the 1960s has downplayed their sometimes monumental failures at bridging the "two cultures," as well as the negative reception they received at the time from both audiences and the press. The models of collaboration put forth by E.A.T, A&T at LACMA, and the Artist Placement Group were all based on the notion of pairing knowledge assets, rather than on a fluid, collaborative exchange that would be of mutual benefit to both scientific and creative discourse, let alone lead to the creation of a third practice that would transcend the limits of the original collaboration. Did these pairings allow for the necessary spontaneity, discovery, open-ended research, and play? Or were the relationships too complex, sometimes with oppositional agendas, and too muddied by capitalism and product-oriented goals?

The institutionalization of artists in scientific or technological environments never happened the way Robert Rauschenberg and Billy Klüver had envisioned when they wrote the first mission statement of E.A.T. in 1967: "The purpose of Experiments in Art and Technology, Inc. is to catalyze the inevitable involvement of industry, technology, and the arts... E.A.T. was founded on the strong belief that an industrially sponsored, effective working relationship between artists and engineers will lead to new possibilities that will benefit society as a whole."

However, in the intervening decades, a change in tone has become evident in the establishment of media centers, the placement of artist-in-residence programs in industrial or scientific environments, and the beginnings of interdisciplinary academic degree programs. These new platforms have helped artists become more hands-on and conversant in scientific and technological methods, rather than employing them from a removed or naive distance. This has as much to do with a change in the way museums and institutions treat this hybrid breed of artistic practice as it does with the technological milieu of our age—people today, artists notwithstanding, have access to resources that simply didn't exist in 1966.

Practitioners now have greater agency to work fluidly across disciplines and beyond rarified institutions and industries. The Internet has provided unprecedented access to knowledge networks, fabrication processes, expertise, and audiences. That is not to say that institutional art/science or art/engineering pairings and artist-in-residence programs have less value today, but rather that they represent just a few possible platforms for such exchanges. Nowadays, networks of artists, scientists, and engineers can be assembled virtually with fewer geographic or economic constraints. From this expanded playing field, the types of activities possible have exploded in number, yielding a variety of methodologies and expressions, which this book attempts to document.

A snapshot of the "now," as this book sets out to take, always presents complications. Categories prove elusive; no guide-books exist to clarify the landscape. Efforts must rely on inadequate and coarse labels. Yet the self-organizing networks of artists, long-distance collaborations, and homebrew tech-nologists documented here all point toward an approaching horizon resembling C.P. Snow's vision. This book is a first draft of present collaborations and crossovers, steeped in historical trends but undeniably a product of today.

Art and science are both manifestations of the human drive for knowledge; they provide their practitioners with a feeling of resonant connection to the complex processes that underlie our environment. And though they ultimately express a different view of the universe, they aren't mutually exclusive—rather, they mirror each other in fantastically interesting ways. We live in a moment of unprecedented change in the way that both art and science are practiced, and those changes are happening in parallel with one another. How we adapt, collaborate, and express our changing environment may ultimately reconcile the "two cultures" and turn us on to a new level of engagement with our hypersensory, interconnected, and evolving world.

1. The Art and Technology (A&T) Program of the Los Angeles County Museum of Art began in 1967 and concluded in 1971. To promote exchanges between artists and the corporate world, nearly forty artists were paired with U.S. companies with the goal of realizing new or technically complex works.
2. Founded in 1966 by Billy Klüver, Fred Waldhauer, Robert Rauschenberg, and Robert Whitman, E.A.T. was a non-profit group, active primarily from the 1960s to the 1980s. Its aim: to mobilize the arts, industry, and science around projects that involved participants from each field. E.A.T. promoted inter-disciplinary collaborations through a program pairing artists and engineers.
3. The Artist Placement Group (APG) emerged in London in the 1960s. The organization actively sought to reposition the role of the artist within a wider social context including government and commerce. APG differed from A&T and E.A.T. in that the product of these exchanges was more theoretical and less aimed at creating physical works of art.
4. General system theory (GST) was defined by Ludwig von Bertalanffy, an Austrian-born biologist, in his 1968 book *General System Theory: Foundations, Development, Applications*, rev. ed. (New York: George Braziller, 1976).
5. C.P. Snow's Rede lecture was given at Cambridge University on May 7, 1959, and led to Snow's subsequent publications, *The Two Cultures and the Scientific Revolution* (Cambridge: Cambridge University Press, 1960), and *The Two Cultures: and A Second Look* (Cambridge: Cambridge University Press, 1964).
6. Pamela Lee, " Eros and Technics and Civilization," in *Chronophobia: On Time in the Art of the 1960s* (Cambridge, Mass.: MIT Press, 2004), p.14.

MAINTAIN A CONSTRUCTIVE CLIMATE FOR THE RECOGNITION OF THE NEW TECHNOLOGY AND THE ARTS BY A CIVILIZED COLLABORATION BETWEEN GROUPS UNREALISTICALLY DEVELOPING IN ISOLATION. ELIMINATE THE SEPARATION OF THE INDIVIDUAL FROM TECHNOLOGICAL CHANGE AND EXPAND AND ENRICH TECHNOLOGY TO GIVE THE INDIVIDUAL VARIETY, PLEASURE AND AVENUES FOR EXPLORATION AND INVOLVEMENT IN CONTEMPORARY LIFE. ENCOURAGE INDUSTRIAL INITIATIVE IN GENERATING ORIGINAL FORETHOUGHT, INSTEAD OF A COMPROMISE IN AFTERMATH, AND PRECIPITATE A MUTUAL AGREEMENT IN ORDER TO AVOID THE WASTE OF A CULTURAL REVOLUTION.

2. Statement, Experiments in Art and Technology, Billy Klüver and Robert Rauschenberg, 1966

PROGRAM ART OR BE PROGRAMMED

REACTIVE, INTERACTIVE, GENERATIVE, COMPUTATIONAL, ROBOTICS, DATA VISUALIZATION

As new technologies emerge with greater rapidity, and as they grow in accessibility, the opportunities for artists to respond by finding alternative, innovative, and expanded possibilities for these recently developed tools increase in kind.

The technologies artists have access to today can be considered the latest in a pantheon of tools that have shaped the face of artistic practice since primitive humans touched pigment to a cave wall. Much has been written about how new media—from the printed word to photography, video, and now hardware and software—go through a period of rejection and redefinition, followed by acceptance into the mainstream art lexicon. We happen to live in an era, however, in which institutions pop up quickly enough to support artists in the interim between these periods of initial rejection and delayed acceptance. Today, organizations like Eyebeam Art and Technology Center, Rhizome.org, transmediale, and Ars Electronica give technologically inclined artists a community, meeting place, and resources.

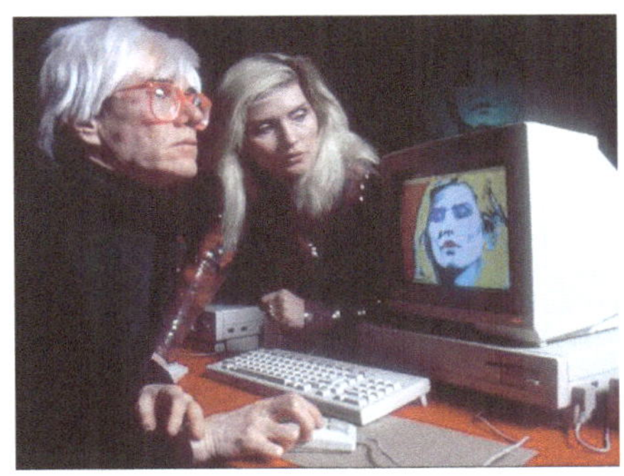

3. Andy Warhol paints Deborah Harry on an Amiga computer at a 1985 Commodore press conference

In the past, artists working with technology were required to adapt unusual tools to their practice. When computer animation pioneer John Whitney wanted to make his works in the 1950s, he had to build his own equipment by converting the directing mechanism of a World War II M-5 anti-aircraft gun into a "cam machine."[1] Whereas Whitney had worked at a Lockheed aircraft factory during World War II and acquired the technical aptitude to realize such undertakings, in other instances artists aspiring to make work that was technological in nature needed to partner with technicians, industrial labs, or engineers. Although the 1960s saw a great deal of collaboration between artists and these unlikely bedfellows—Experiments in Art and Technology provided pairings of artists and engineers, while early computer animator Larry Cuba created his work "after hours" using downtime on the computers at the Jet Propulsion Lab—the scarcity and cost of tools precluded independent artists from achieving an abundance of work in this domain. Indeed, this was the case for a good part of the twentieth century. David Hockney using a Quantel paintbox for a BBC special in 1989 and Andy Warhol painting Deborah Harry with a Commodore Amiga in 1985 were rare enough occurrences that they warranted media coverage.

The introduction of the personal computer in the 1970s, and its exponential reduction in cost over the last four decades, shattered this precedent. A new breed of artist has emerged, capable of making work of tremendous scope on computers that would have been prohibitively expensive just a decade ago. The delivery method of the Internet gives such work the potential to be seen by millions. The last ten years have seen a radicial shift in the perception of computer technology, from something that is exclusively developed by industry professionals to something that is an extension of DIY culture. Open source hardware and software communities, like openFrameworks[2] and Arduino[3] developers, have helped make digital tools accessible and customizable by artists and non-artists alike.

1. A "cam machine" is an analog computerized motion camera.
2. openFrameworks is an open source toolkit designed for "creative coding."
3. Arduino is an open source single-board microcontroller, designed to make the process of using electronics in multidisciplinary projects more accessible.

In regard to the art historical lineage of this work, the artist and software programmer C.E.B. Reas holds that software, like thought, is immaterial, and he sees a clear relationship between the "conceptual" art of the 1960s, namely Minimalism, Conceptual Art, Op Art, and Fluxus, and the "computational" art today. He writes, "Software has enabled a way to build a bridge between the art of the past and the electronic arts of the present and future."[4] It could further be argued that computational art is directly descended from the "systems" art[5] of the 1960s, which was less concerned with material art objects and more with systems and processes.

Knowledge of the mostly invisible workings of programmable devices has a practical value, too. The media theorist and critic Douglas Rushkoff has an axiom: "Program or be programmed." In a talk given at the SXSW Interactive Conference in 2010, he argued, "If we don't create a society that at least knows there's a thing called programming, then we will end up being not the programmers, but the users—and, worse, the used."[6] Whether or not the artists and designers using this methodology are interested in unveiling and commenting on the social, cultural, or ethical consequences of technology, the act of learning and sharing techniques that most people accept passively is a statement of emancipation from unidirectional technological consumption.

4. C.E.B. Reas, "Programming Media," available at http://reas.com/
5. "Systems esthetics" was popularized by the author Jack Burnham in his 1968 publication *Beyond Modern Sculpture: The Effects of Science and Technology on the Sculpture of This Century* (New York: George Braziller; London: Allen Lane/Penguin Press).
6. Douglas Rushkoff, "SXSW 2010: Program or be Programmed: Ten Commands for a Digital Age," available at http://youtu.be/imV3p-PIUy1k

C.E.B. REAS
PROCESS COMPENDIUM (A)
2004-2010

C.E.B. Reas, or Casey Reas, is an artist and software programmer who uses advanced programming techniques to make his artwork. With Ben Fry, he created the open source programming language Processing while he was a graduate student at the MIT Media Lab. Processing, which can be implemented for anything from text manipulation to image rendering and video, is currently used by thousands of artists and designers worldwide as an "electronic sketchbook" from which they generate finished works of varying dimensions. It has revolutionized the digital arts environment by placing artists in a direct relationship with programming code. Reas himself uses his software (both Processing and the original code) to translate documents from natural language into visual representations—videos and still images that serve as visual renderings of semantic ideas.

Reas, who cites John Cage and Sol LeWitt as influences, is interested in emergence: he writes that the most important element in his work is the natural language text with which he begins. This text establishes simple relationships among the elements of the piece by specifying behaviors, such as "when touching another, change direction," or "constant linear motion." The software elements, as Reas composes them, comprise both these behaviors and a visual form. Once the software is set into action, Reas observes how the behaviors emerge into a web of relationships. This implementation of the software is the finished piece, and though the result can be represented to the viewer in a number of different forms—print, software, installation, video—none alone reveals the full complexity of the work.

According to Reas, "Each new work is called a Process and defines rules and instructions which describe a relation between Elements outside of a specific physical media."[1] By working with software as his primary medium, Reas has the ability to engage "live" processes, altering parameters and essentially tinkering with the physics of his own design, then by "freezing" the processes into printed images—essentially plucking one image from an infinite number of possible variations—he can observe the output of his work in greater detail. Indeed, Reas himself may be the only person to really understand the kinetic activity underfoot in each print.

Organic forms rise unpredictably out of Reas's machine language with an intensity of aesthetic beauty that is almost alchemical. In *The World of Digital Art*, Mitchell Whitelaw writes that Processing-based work such as Reas's is "both abstract and concrete, formal and sensual, technological and cultural."[2]

1. C.E.B. Reas, "Process / Drawing," available at http://reas.com
2. Mitchell Whitelaw, quoted in Wolf Lieser, ed. *The World of Digital Art* (Potsdam: h.f.ullmann publishing, 2010).

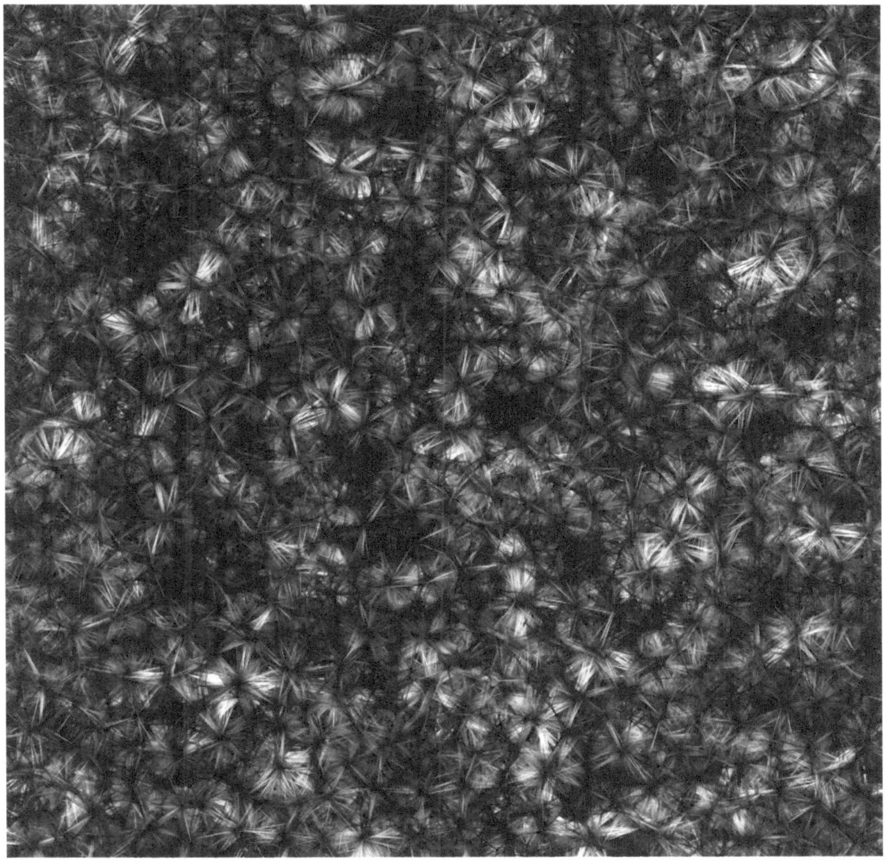

4. C.E.B. Reas, *Process 4 (A)* from Process Compendium 2004-2010

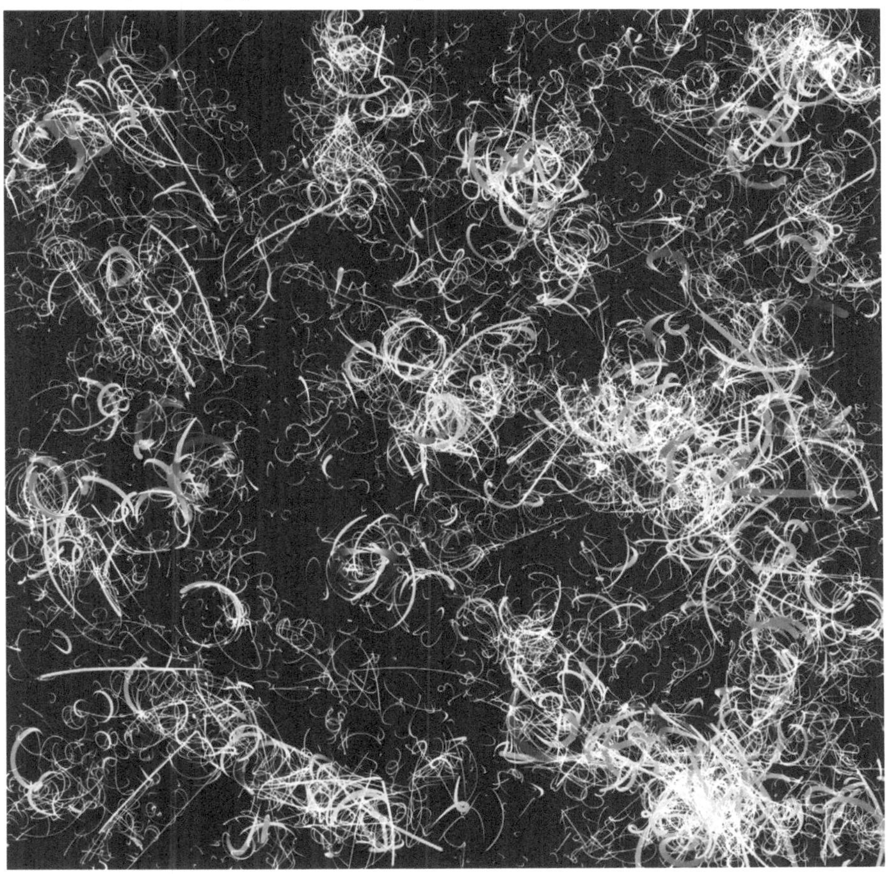

5. C.E.B. Reas, *Process 11 (A)* from Process Compendium 2004-2010

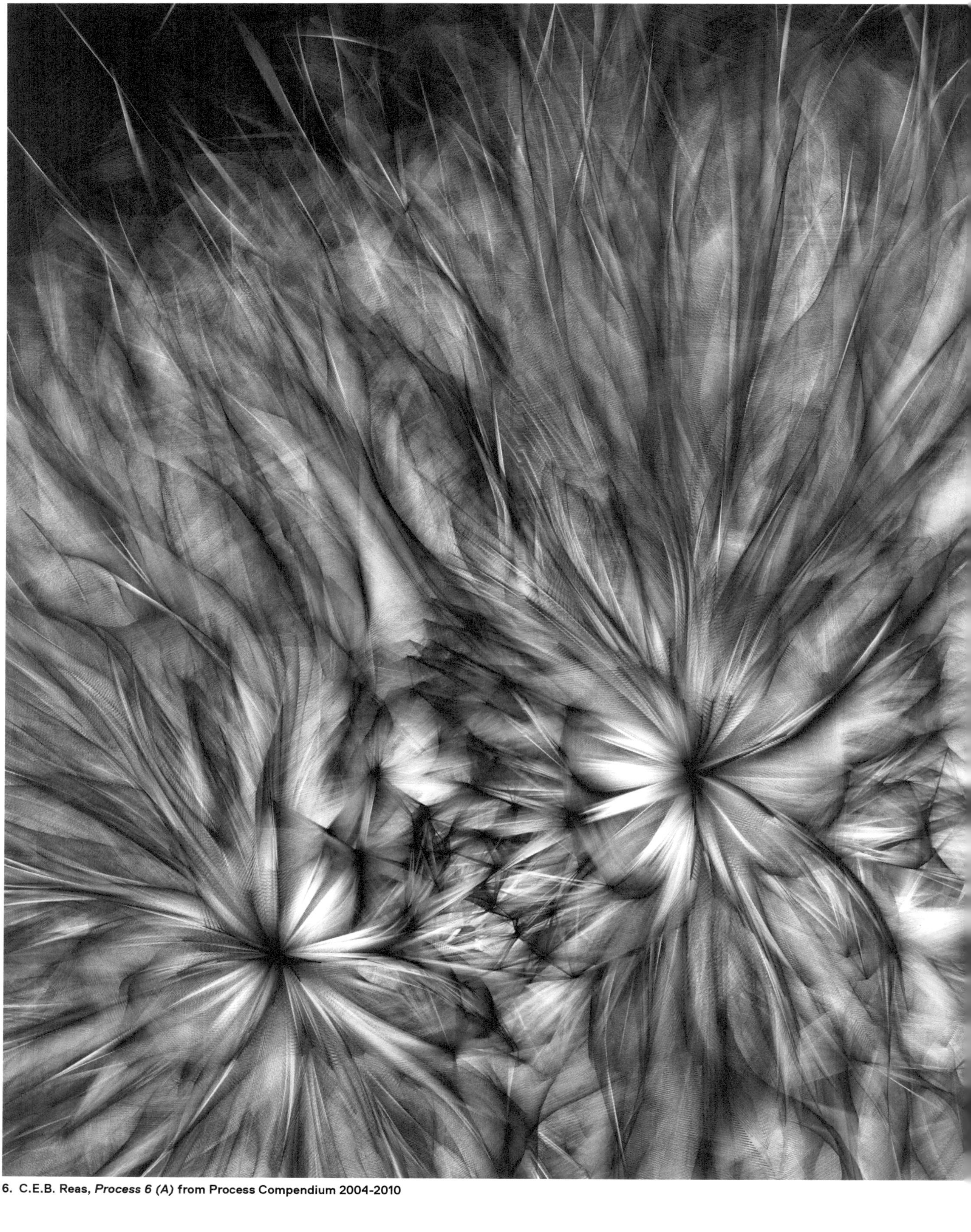

6. C.E.B. Reas, *Process 6 (A)* from Process Compendium 2004-2010

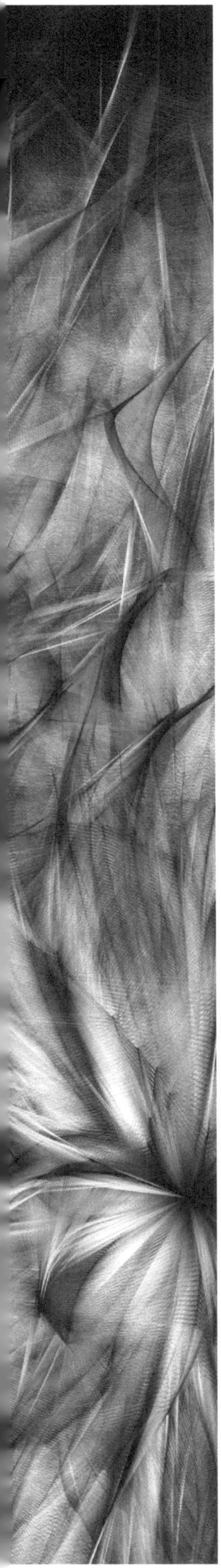

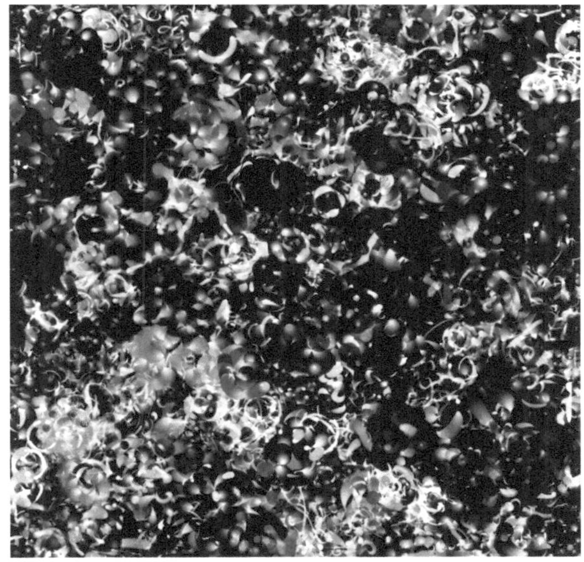

7. C.E.B. Reas, *Process 14 (A)* from Process Compendium 2004-2010

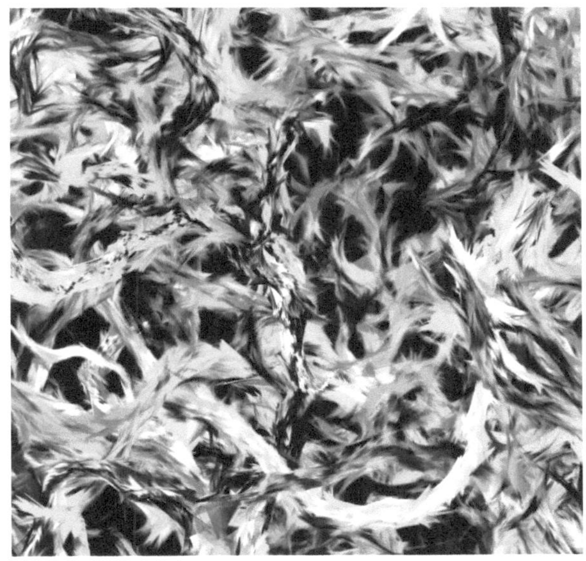

8. C.E.B. Reas, *Process 18 (A)* from Process Compendium 2004-2010

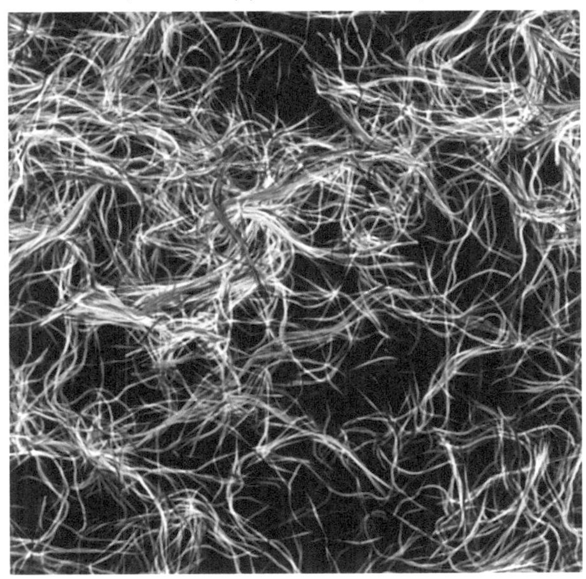

9. C.E.B. Reas, *Process 17 (A)* from Process Compendium 2004-2010

RAFAEL LOZANO-HEMMER
PULSE ROOM / PLEASE EMPTY YOUR POCKETS
2006-2010

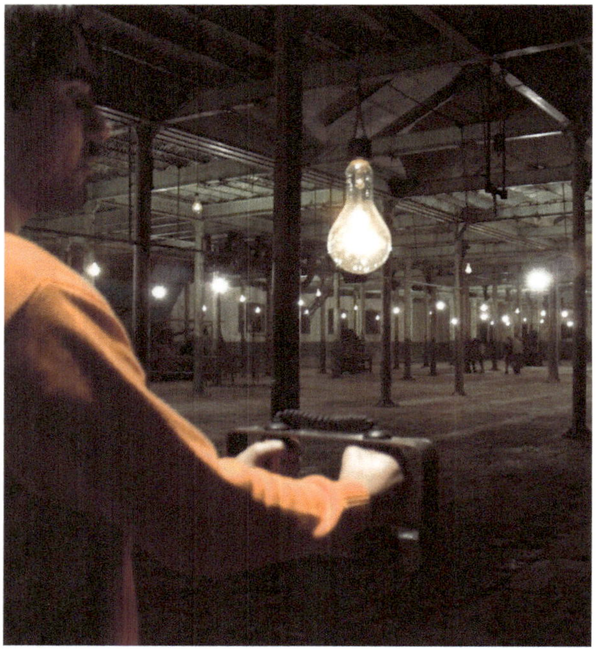

10. Rafael Lozano-Hemmer, *Pulse Room*, 2006

Rafael Lozano-Hemmer is a Mexican-Canadian artist with a degree in physical chemistry and a practice that involves architecture, technological theater, and performance. Widely known for creating large-scale interactive installations in public spaces around the world—his *Vectorial Elevation* (1999-2000), comprising eighteen robot-controlled searchlights mounted on public buildings and rigged to perform "choreographies" based on participants' online suggestions, won several awards including the prestigious Golden Nica at the 2000 Prix Ars Electronica—Lozano-Hemmer makes work that uses technology to prompt viewer interaction.

In his current output, he has scaled down the grandeur of his earlier pieces into playful "shadow boxes" and "subsculptures" that present the viewer with a relational experience—a direct engagement with both the legacy and immediacy of technology. He has moved from the macrocosm to the microcosm, from public squares and building facades to fingertips and pockets. In 2010's *Please Empty Your Pockets*, viewers are prompted to place whatever small objects they may have on their persons (lighters, pens, keys) onto a conveyor belt, where they pass through a computerized scanner that "prints" the objects onto the belt. The piece remembers up to 600,000 objects, displaying them alongside the new images that are perpetually being added to the installation. The piece, Lozano-Hemmer writes, "intends to blend presence and absence using traditional techniques of augmented reality."

Other pieces, interactive displays and installations with built-in computerized tracking systems, engage their audience in similarly tactile ways. *Pulse Room* (2006) asks the viewer to hold a biometric sensor, which immediately sets off one of three hundred light bulbs hung from the room's ceiling to pulse in synch with the viewer's heartbeat. The moment the sensor is released, the flashing sequence passes on to a different bulb, allowing the next participant to imprint a heartbeat on the first light bulb. At any given moment, the installation holds the recordings from all of its most recent participants; the end result is a room of lights, humming and flashing with human echoes.

Lozano-Hemmer's work, which necessitates the viewer's direct tactile engagement, speaks to our escalating intimacy with technology.

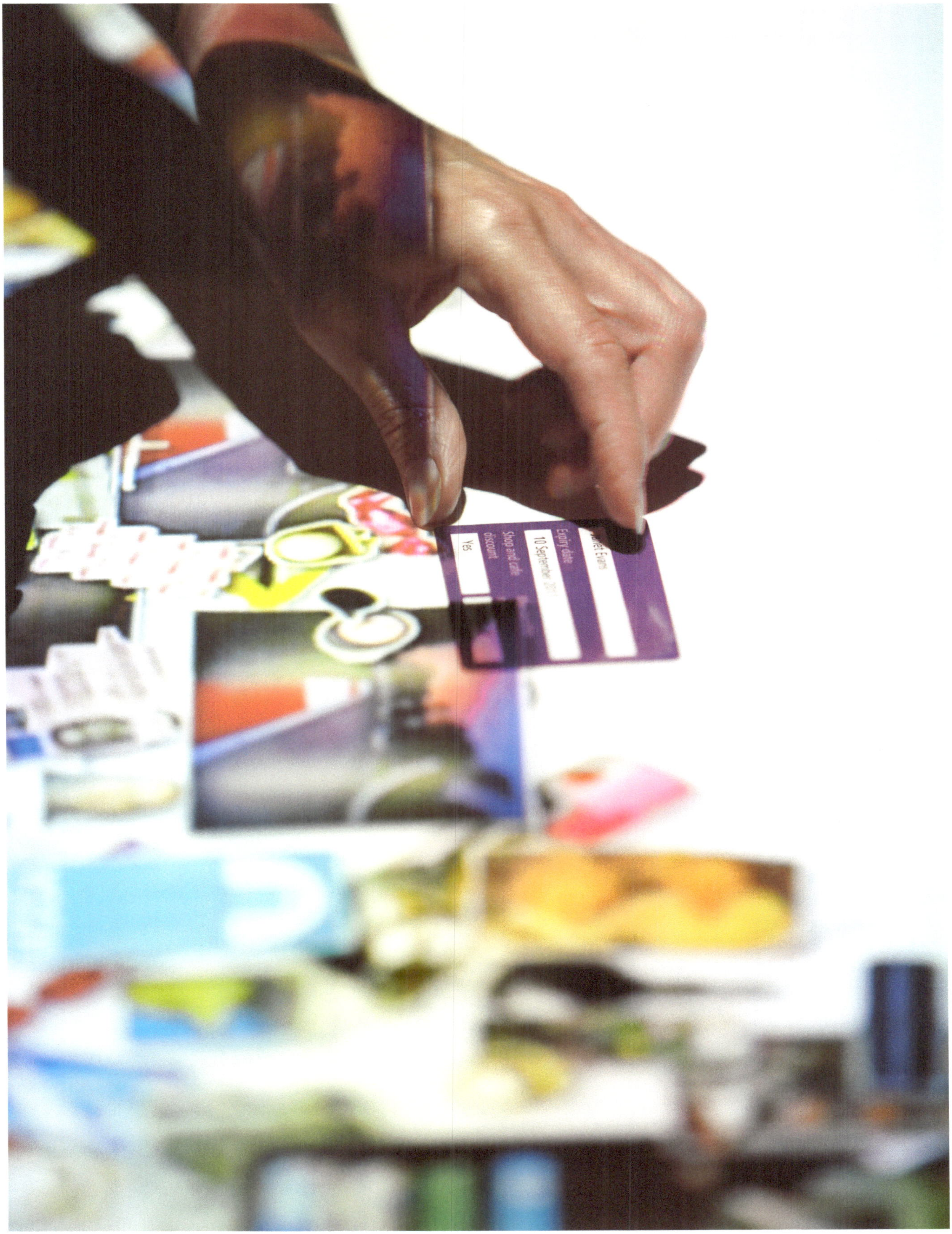

11. Rafael Lozano-Hemmer, *Please Empty Your Pockets*, 2010

JER THORP
NEW YORK TIMES 365/360
2009

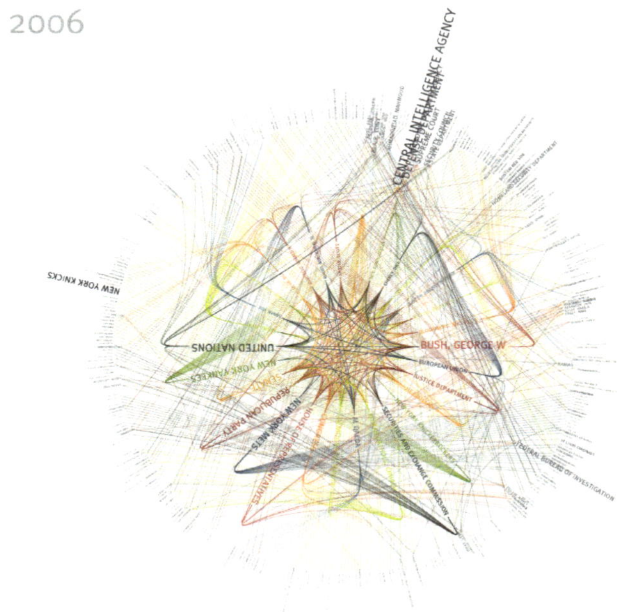

12. Jer Thorp, *New York Times 365/360*, 2006

Jer Thorp is an artist from Vancouver, Canada, whose practice—software-based work as well as the occasional hardware piece—conflates science, art, and daily life. He is known for making visualizations, using the Processing programming toolkit, that displays data in powerfully aesthetic ways: when the *New York Times* released its Article Search API (Application Programmer Interface), Thorp used it to make elegantly comprehensive images charting the relationships between, say, the frequency of the words *hope* and *crisis*, or *sex* and *scandal*, in the newspaper's history. In his *New York Times 365/360* pieces, Thorp uses Processing to diagram the most frequently mentioned people and organizations of every year dating back to 1985. The result is a highly complex digital image that includes a staggering amount of information.

Like C.E.B. Reas, Thorp considers the real work to be his software, "not the charts that come out of these things, but the actual programs that I distribute into the world." Thorp makes the source code for most of his projects available for free on his website, with the intention of allowing other artists to build on what he has made.

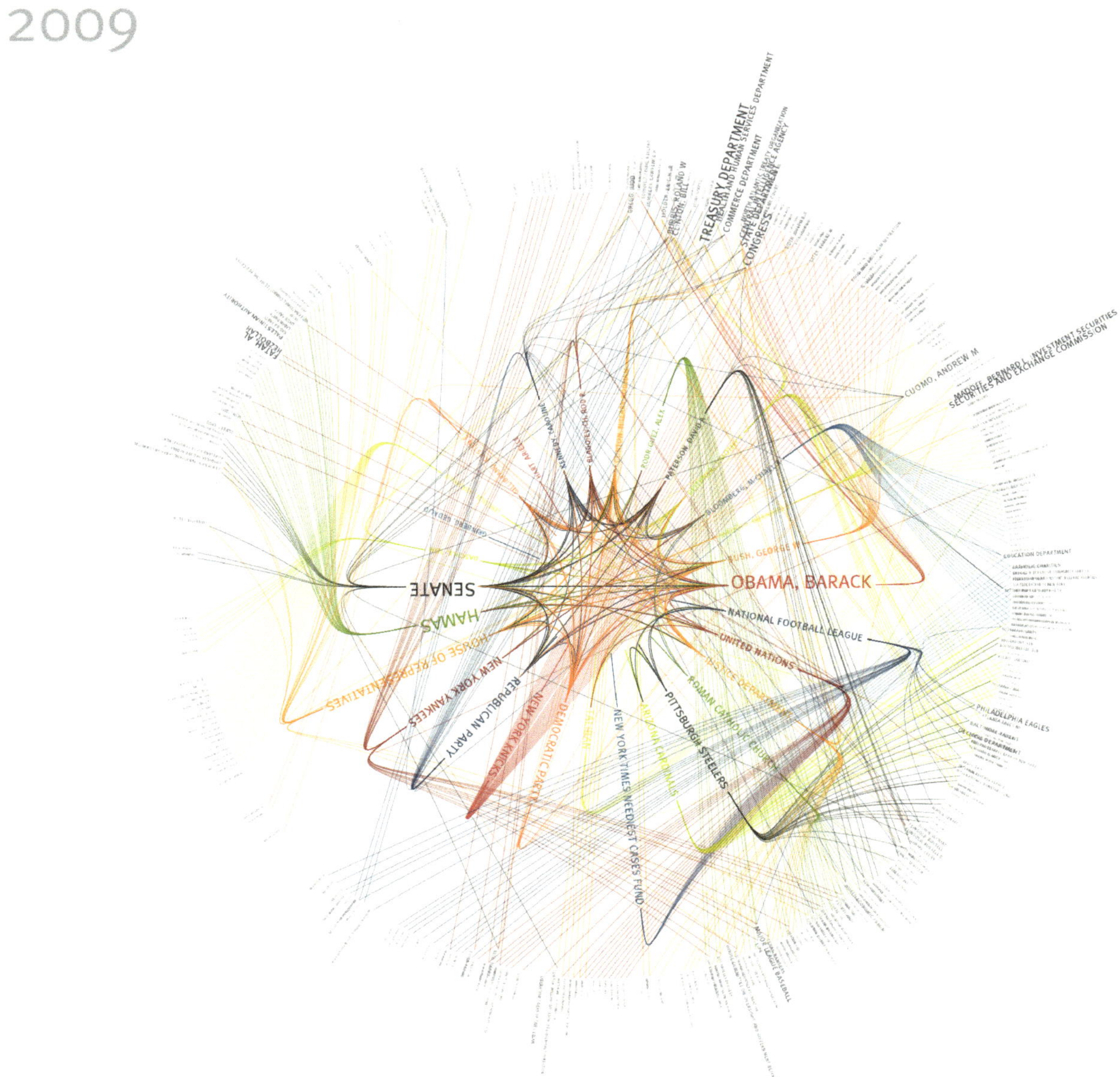

13. Jer Thorp, *New York Times 365/360*, 2009

Golan Levin

As an artist, how do you see your role in a technological or scientific setting?

Role: my art is a form of research into what I call "speculative human-computer interaction design".

My purpose within a technologically-oriented institution is:
— To expand the vocabulary of human action by means of arts-based approaches to innovation
— To ensure the presence of a humanist and critical perspective in the pursuit of technological 'progress'

This comes down to participating in interdisciplinary design teams by:
— proposing novel and unexpected problems for consideration
— contributing to brainstorming sessions
— anticipating the implications of, and devising provocative applications for, new technologies

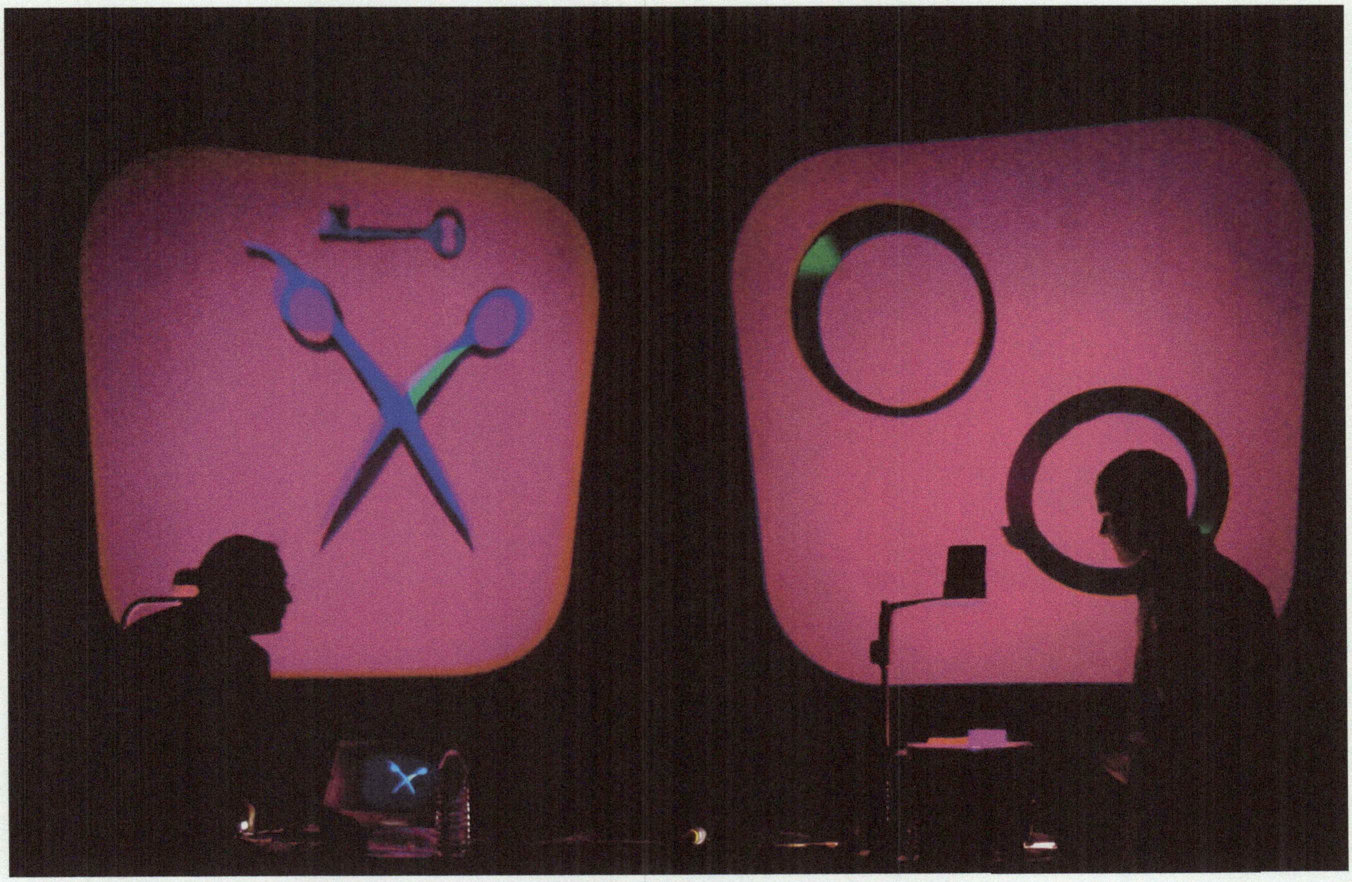

14. Tmema (Golan Levin and Zachary Lieberman), The Manual Input Sessions, 2004

MARIUS WATZ
STOCKSPACE AND OBJECT #1-3
2006-2009

Marius Watz is an artist working with visual abstraction through generative software processes. Like C.E.B. Reas, he uses the Processing programming toolkit to synthesize forms and create parametric behaviors, but in Watz's case, he has made the leap from the screen to the physical by implementing digital fabrication techniques. Using three-dimensional printers, laser cutting, and CNC (Computer Numerical Control) milling, Watz transforms his software, and the purely digital network of relationships and behaviors it creates, into objects.

Watz explains the evolution this way: "Since the 'original' is a piece of digital information rather than a physical mold, there is no reason why the model should not be a piece of dynamic software, capable of responding to parametric control. Why not allow users to co-design their product to their own specification through a software interface, then produce it on demand?"[1]

For Watz this return to tactility is an expression of the digital artist's quest for the ultimate "high resolution" object. After all, although digital processing can now shoulder prodigious amounts of information, a physical thing for the human perceiver is still the most substantial representation of information possible. Unlike something on a screen, a real object requires no suspension of disbelief. Software-made objects are a new category: the finished product is unlike anything that could ever be molded or sculpted by hand, and yet it's distinctly the product of human creativity.

1. Marius Watz, quoted in Wolf Lieser, ed., *The World of Digital Art*. (Potsdam: h.f.ullmann publishing, 2010).

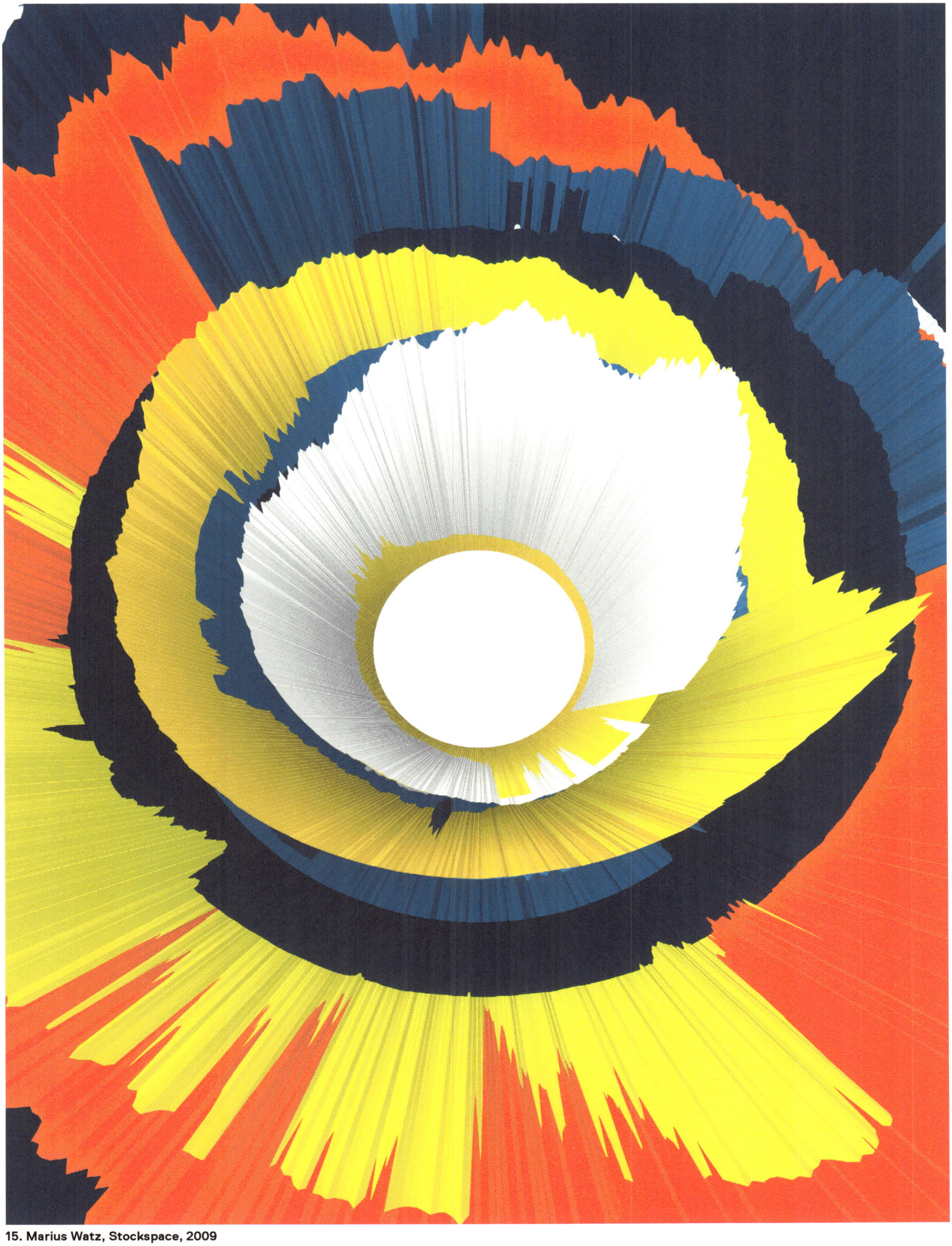

15. Marius Watz, Stockspace, 2009

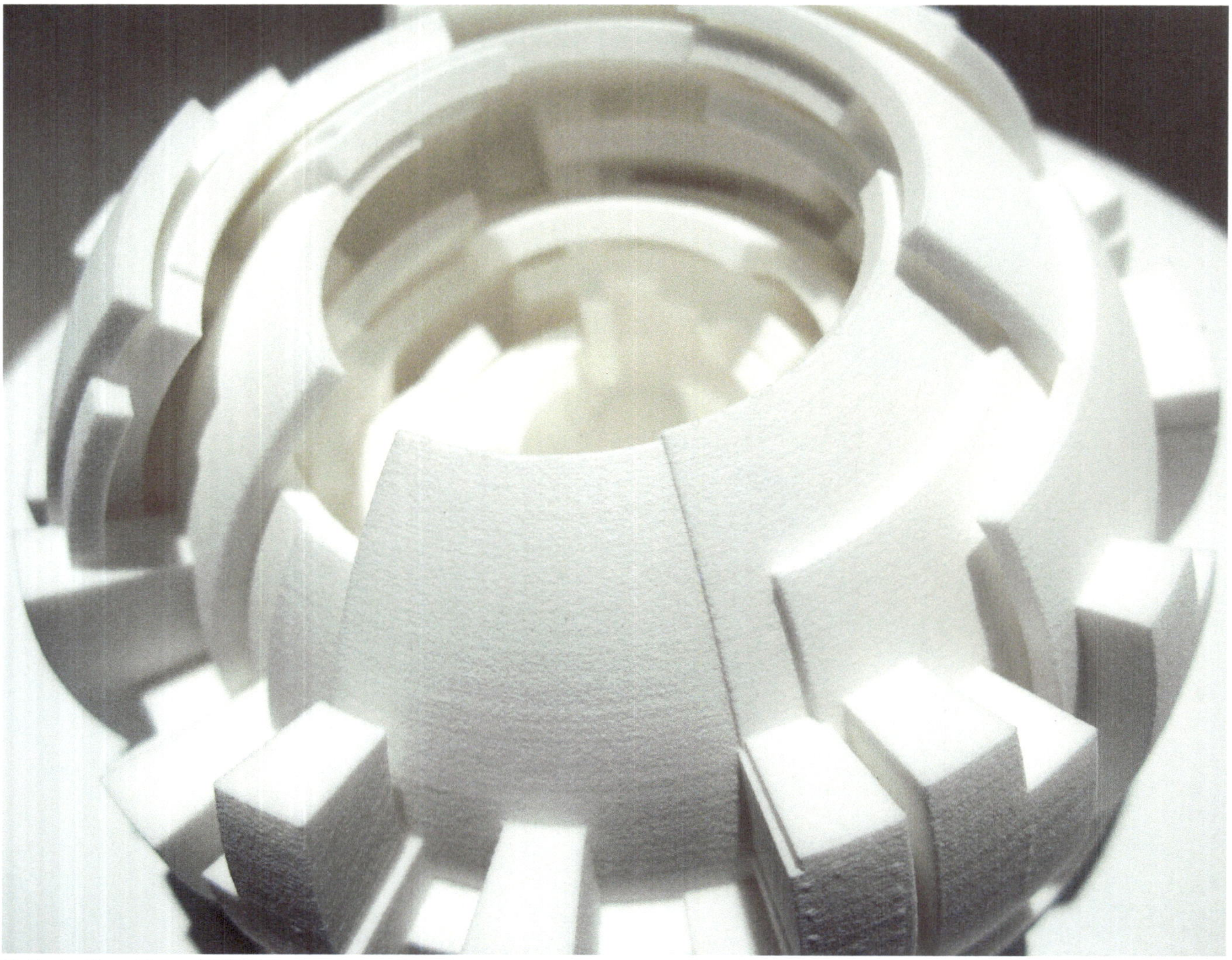

16. Marius Watz, *Object #3*, 2006. Image courtesy the artist

AARON KOBLIN
BICYCLE BUILT FOR TWO-THOUSAND
2009

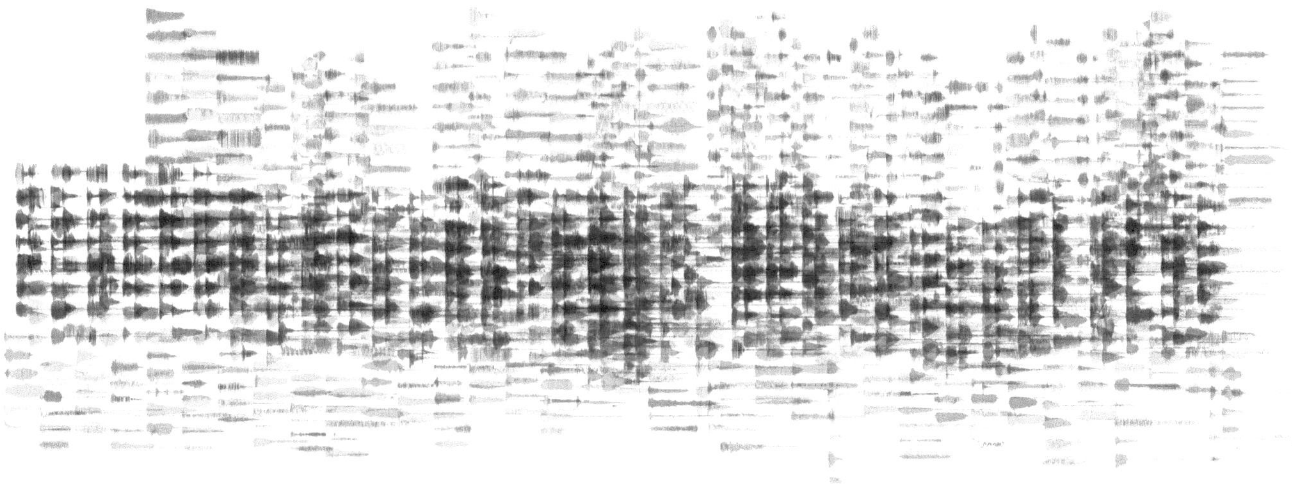

17. Aaron Koblin, *Bicycle Built for 2000*, 2009

Aaron Koblin was one of the earliest artists to realize the potential of utilizing online "immaterial labor" to create works of art with potentially thousands of virtual contributors. He is known for "crowdsourcing" via the Amazon Mechanical Turk service—a kind of Internet bazaar of human intelligence—to make digital works created by anonymous networked contributors. In 2006, he enlisted over 7,000 Mechanical Turks to "draw a sheep facing to the left" for two cents each. The result was 10,000 hand-drawn sheep, which were presented as an animated installation and sold as physical prints. Participants were unaware of the end product of their labor: a work of art titled *The Sheep Market*. In this way, Koblin began to tease out the nuances of what it means to inhabit the online marketplace, and how one's labor or ideas might be unknowingly used for monetary gain.

In 2009, Koblin took the idea of crowdsourced art further with *Bicycle Built for 2000*, in which 2,088 people were asked to mimic an individually provided tone, actually tiny fragments of the song "Daisy Bell;" their voices were then stitched together to reconstruct the song. This is a song historically performed by computers: in 1961, an IBM 7094 at Bell Labs was programmed to perform the song, becoming the first computer to sing, and later, of course, it gained notoriety as the dying song of the artificial intelligence HAL 9000 in *2001: A Space Odyssey*. The combination of a mechanical voice with this affably romantic ditty has proved to be a lasting (and eerie) juxtaposition. The end result of Koblin's *Bicycle Built for 2000*, a drone of human voices that sounds like an army of robots, is a coy inversion of this pop culture image.

Aug. 3, 1965 R. B. FULLER 3,197,927

GEODESIC STRUCTURES

Filed Dec. 19, 1961 6 Sheets-Sheet 3

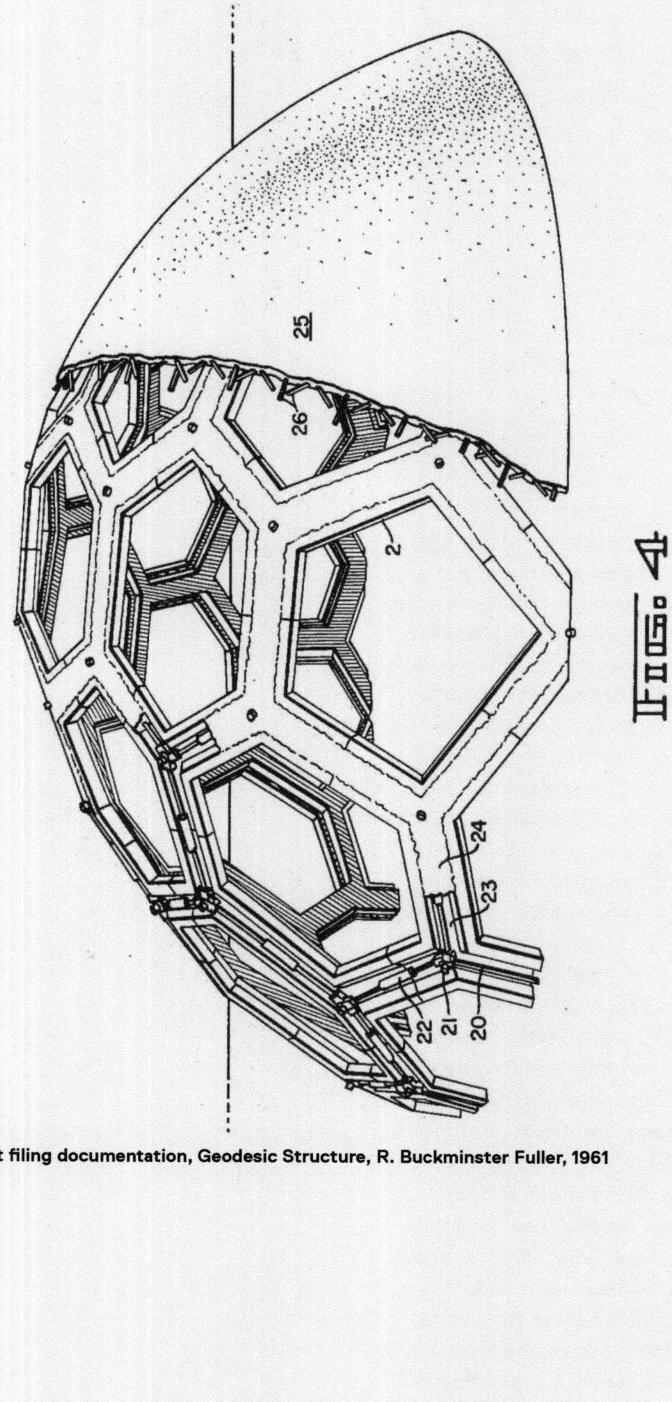

18. Patent filing documentation, Geodesic Structure, R. Buckminster Fuller, 1961

SUBVERT!

HACKER, HACKERSPACES, HACKTIVISM, TINKERING, JAILBREAKING, TACTICAL MEDIA

The term *hacker* has been so corrupted by misuse in the past twenty years that to call contemporary co-opting or repurposing "hacking" devalues and mislabels the varied creative methodologies at work today. To more accurately discuss the work of hacker artists, consider the definition of *hacking* from a primary source: *The Jargon File*, a hacking glossary, attributes various characteristics to hackers, most notably: "One who enjoys the challenge of creatively overcoming or circumventing limitations." While the uninitiated associate criminal behavior with hacking, *The Jargon File* corrects this misconception by labeling programming malfeasance as the work of *crackers*.

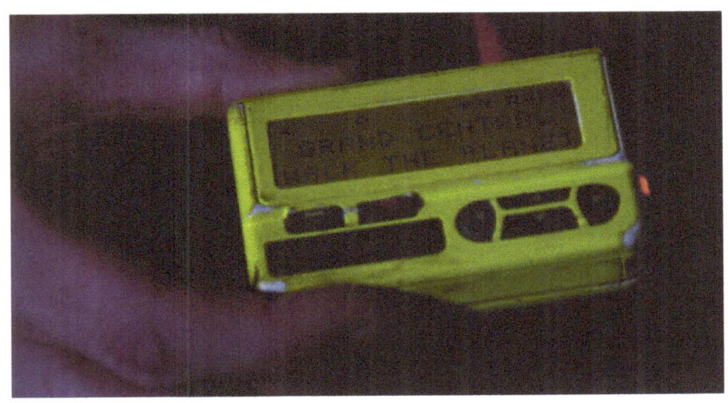

19. Image from the 1995 film *Hackers*

Today, those who assume the title of "hacker" in its creative spirit are not exclusively members of the programming subculture. Contemporary hackers have a strong interest in the way things work; they like to tinker, customize, modify, and repurpose existing and obsolete technologies, and as a group they tend to embody the altruistic principles of collaboration and information sharing. That is not to say that hacker activities are free from pranks, political motivations, and anarchistic impulses. But in general these sorts of activities are directed less toward doing harm and more toward freedom in the broadest sense—freedom from limitations imposed on speech, the use of manufactured goods, access to information, and personal expression. The community-minded side of hacking is demonstrated by the current international network of "hackerspaces" that tend to resemble a shared studio or community machine shop more than an underground hideout for criminal activities.

Many of the operating procedures of contemporary hacker artists are descended from the acts of appropriation in twentieth-century art. The story of modern art could be told as the conceptual shift from perceptual goals (creating representative and "realistic" images) to self-reflective and experimental models. Marcel Duchamp turned a urinal into a "fountain." Pablo Picasso included an image of chair caning in a collage. Robert Rauschenberg made a "combine," while Andy Warhol and Roy Lichtenstein problematized the relationship between copies and the original. These referential methods could be perceived as a direct precedent for current hacker works, but the combination of this legacy and today's hacker culture has produced an appropriation paradigm that is neither geek nor modern.

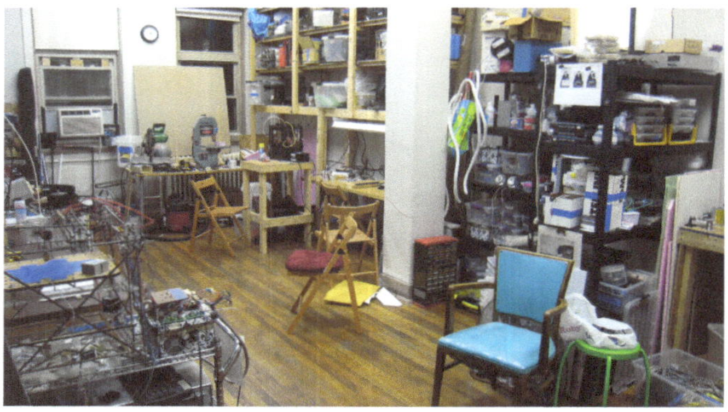

20. HacDC, a hackerspace. Photo: Patrick and Preston Thomas

Rather than rule-breaking, contemporary hackers pursue rule-*bending*. What hacker work is *about*—the structures and limitations under which it operates—is primarily a concern with things like cultural norms, commercial products, consumer technology, legal standards, and geopolitical conditions.

Once the limits of these things are identified, many hackers use those limits to toy with, or augment expectations of, the normal operating procedure of current systems. The familiar or "normal" state of things is manipulated. The expected is made unexpected. Loopholes are found. But the object being hacked remains recognizable in order for the grace of the hack to be appreciated. In brief, the "rules" *are* the medium within which hacker artists swim. They produce unexpected results that would expand rather than reduce limits.

The works we cite here are examples of subversion through an inverse (and at times perverse) attitude about technology. Rather than combat the consequences of technology with non-technology, today's hackers dive headlong into technology. They disrupt the mechanisms of everyday life, repurpose consumer technology, and unravel the nature of digitality. These inversions of former subversive tactics announce the powerful influence geek hackers have had on contemporary art.

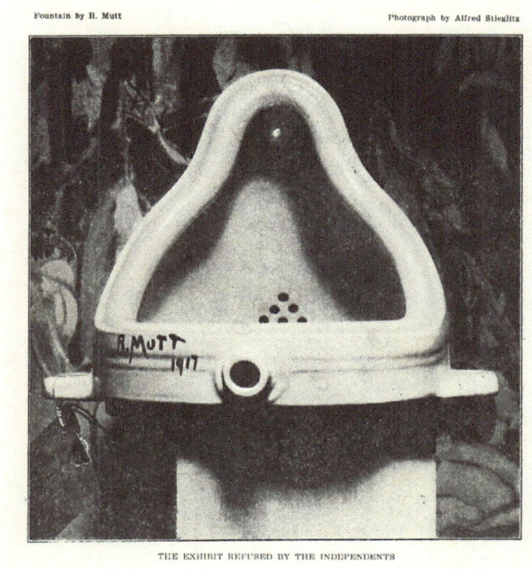

21. Marcel Duchamp, *Fountain*, 1917

ROBIN HEWLETT AND BEN KINSLEY STREET WITH A VIEW 2008

Street With A View addresses the tension between surveillance concerns and the triviality of most of the images captured by Google Street View for its mapping system. The taking of such photographs has given rise to debates about privacy and the right to publish and use for commercial purposes the images of individuals and of entire neighborhoods.

With the complicity of both the inhabitants of Sampsonia Way in Pittsburgh and Google Street View, artists Ben Kinsley and Robin Hewlett staged collective performances and actions that took place just as the Google Car was driving through the neighborhood, including a seventeenth-century sword fight, a lady escaping through a window using bed sheets, a gigantic chicken, a parade with a brass band and majorettes, the lab of the inventor of a laser that makes people fall in love, etc. The images that document the events have become an integral part of the Google image archive.

As images cannot replace direct physical experience, they always constitute a reconstruction, if not indeed manipulation, of the real world, but one that we are led to regard as real in today's media-driven society. According to Paul Virilio, the representation of reality in an image then becomes a reality in turn, but reality of lower degree. As the image increasingly displaces the word, it has become the contemporary language most used and of greatest importance in conveying "truth" in the globalized world.

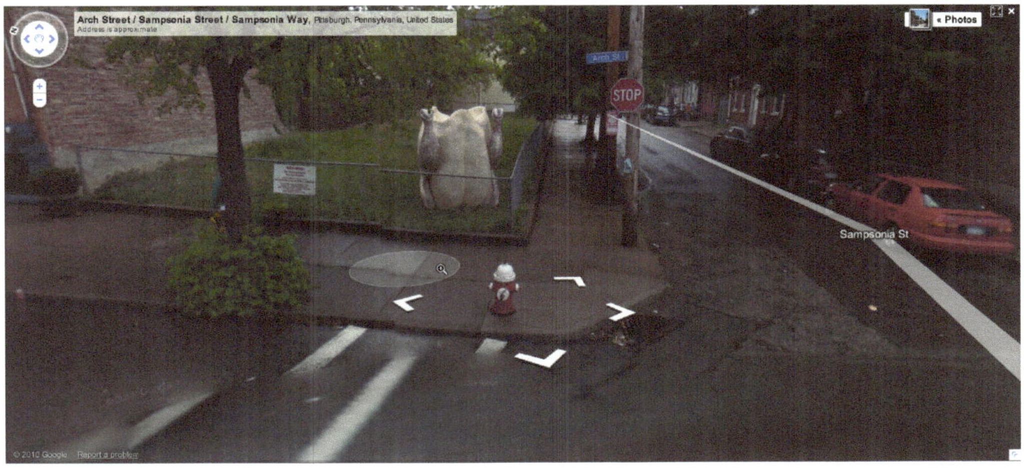

22.

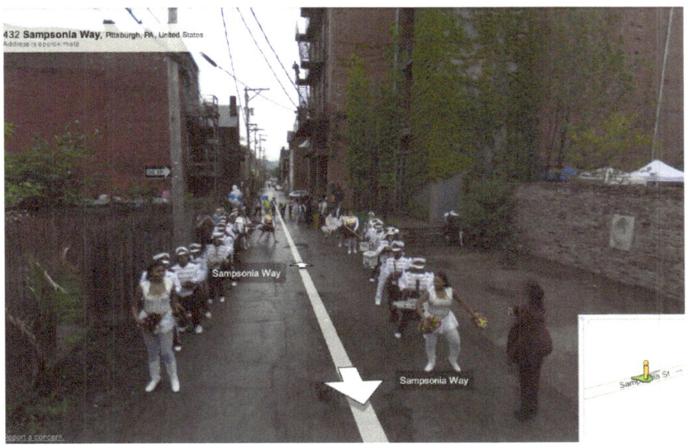

23.

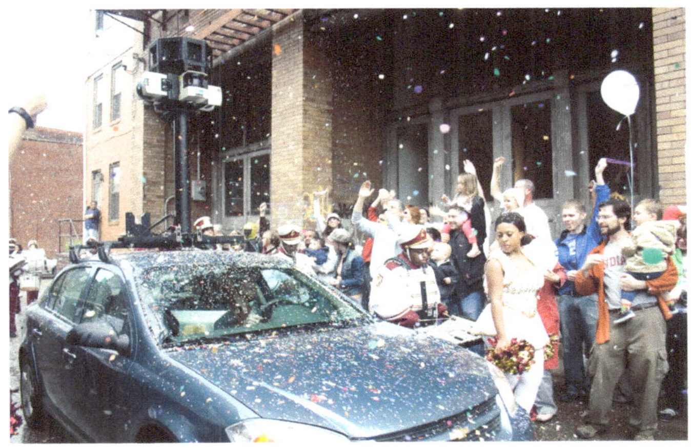

24.

22–24. Ben Kinsley and Robin Hewlett, Video stills from *Street With A View*, 2008

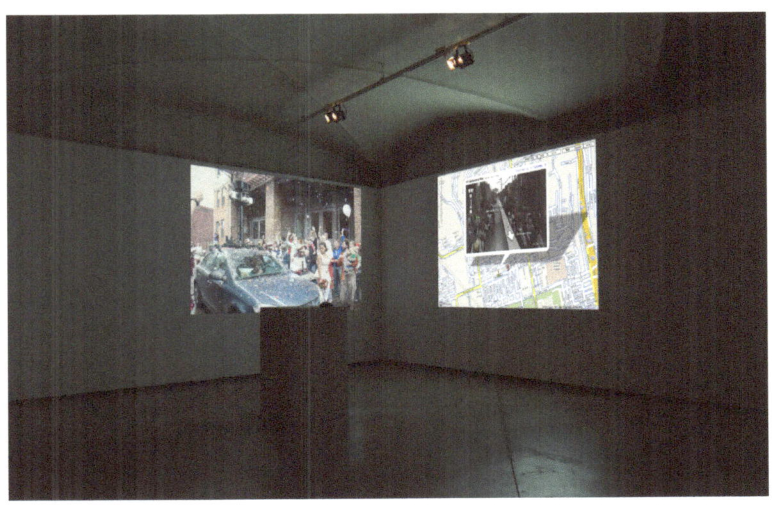

25. Ben Kinsley and Robin Hewlett, *Street With A View*, installation view at Centro di Cultura Contemporanea Strozzina, Firenze, Italy, 2008. Photo: Valentina Muscedra

Skype interview

Johannes Grenzfurthner

Founder and Artistic Director of monochrom, Vienna
Sunday, February 20, 2011

RÉGINE DEBATTY: I am going to explain what we are doing.

JOHANNES GRENZFURTHNER: Okay.

RD: We are doing this book sprint, which tries to cover what artists are doing with technology and science and the possible overlaps. One of the chapters is about hacking, so we immediately thought, "Oh! Let's have monochrom[1]!" Only thing is, I had a look at the monochrom website, and I could not choose a project — actually none of us could choose a project. So I thought maybe we should play it differently and use monochrom as what we call a "Wild Card." In the book, there is an introduction about the program, the hacker, the makers, the researchers, and um... and yeah... I don't remember what I wanted to say...

JG: [laughs]

RD: But instead of having just a project of monochrom's with a short description, like we did with the other artists whose work we present in the book, we thought we could try to interview someone from monochrom and insert it into the hacker section.

JG: So the chapter is about art, hacking, and technology.

RD: Yeah, mostly about that. There is going to be Julian Oliver and Danja Vasiliev with Newstweek.[2] Then there are some projects that are a bit more design-y and extend the word hacker, um... That is actually going to be my first question.

JG: Okay.

1. http://www.monochrom.at/
2. http://newstweek.com/

RD: I had the feeling that hacking used to be seen as a rather dark and threatening activity. In the past people had a very limited view of what hacking was and had a bad opinion of hackers. Nowadays everyone is somehow "hacking." It has become one of the cool words of the moment and it is applied in almost any kind of context. Do you feel that the term hacker has been diluted or maybe has been evolving over time?

JG: I think it has definitely evolved over time. I think that the definition of hacker was the definition given by the media, though I think most hackers didn't actually know what they were until the media started calling them hackers and they kind of liked it. They also liked that dark aura of being seen as the bad guys, the bad asses, the bad ones who break into computer systems

and stuff like that, because it's a nerdy thing. So in the 1980s being a hacker, being called a hacker, was actually something like peer group knighthood. It meant that the media was paying attention to [us], calling our scene "hackers," so we could feel interesting. In the 1990s there was a drastic decline in people who called themselves hackers because hackers lost their status of distinction. In the 1990s when the Internet arrived in the realm of cultural technology, the hackers cried out in pain: "Oh, my god! What are we going to do now that everyone has an Internet account? What is our place here?" It was one of those moments when a counterculture ended up as an over-the-counterculture.

Last year I was in HOPE [Hackers On Planet Earth],[3] a big hacker conference in New York. It was right at the moment when the whole WikiLeaks thing took off, and it was interesting to see how everyone there felt so great, because being a hacker suddenly didn't feel anachronistic anymore. I can make parallels with being a digital artist: many of the apps I have on my iPhone are much more interesting than most of the works I can see at transmediale.[4] So there is always this point when a wannabe avant-garde movement, whether it is hacking or digital art, appears obsolete because the mainstream is now doing what all those guys were doing in the 1980s and 1990s. But last year, suddenly WikiLeaks was the big thing: "Oh, my god! The Department of Defense is really interested, blah blah blah, so we can feel important and great and the elite again." It was quite interesting to see that.

The second thing that happened—and it's actually a really interesting thing because it's most interesting for hackers—is that within the last five years, the so-called hackers space movement took off. So small tinkering and cool working kinds of spaces were being started all over the globe. Of course, there were a couple of hacker spaces already in the 1970s and in the 1980s, especially in the Netherlands, emerging out of the squatters' scene. Then in the 1990s a couple of more places, like c-base[5] in Berlin, emerged. The past five years, however, have seen a Cambrian explosion of hacker spaces. There were [once] maybe fifty of them on the planet, and now there must be between five hundred to one thousand hacker spaces spread everywhere.

The interesting aspect about that is that the term, the definition of hacking, is, of course, broader, because it is about making, it's about meeting, and it's about do-it-yourself—and not only about networking and computer security. Some hardcore hackers, of course, say, "No, that's not hacking, and we don't like that it is being called hacking." But then again the term hacking itself goes back to MIT when students used the term hacking for "pranking." From the beginning there has never been a clear definition of what is hacking and what is not. Of course, there were jargon files and stuff like that, where people tried to define "that is hacking" and "that is not hacking," but I guess that the definition that is accepted right now is that of "using technology in a way that it should not do."

RD: [sound of confirmation]

JG: That you use technology in a way that technology is not intended to be used. And you play with it, and you open it, and you are experimenting with it, and that's, I guess, hacking. So in a certain way, knitting is a form of hacking, too.

RD: [laughs]

JG: It's just defined [in accord] with what you would want to do. I mean, we at monochrom, we are a political group and don't care about terms. For example, we like calling ourselves artists because it's a tactical term. We get money

3. http://thenexthope.org/
4. http://www.transmediale.de/
5. http://www.c-base.org/

for it and sometimes it helps with not being arrested. You can call something "art" and you can get away with it. However, if you would call that same work "activism," you might run the risk of being sent to jail. Labeling yourself artists allows you to do things that you couldn't do otherwise. And I think it's the same thing for hacking. People use it as a tactical term for this and that, and I think you can talk to a million people and you would get a million different definitions, but the core element is doing things that you should not do with technology.

RD: You even answered my second question, which was about how you would define monochrom. And, yes, it seems to be hacker, activist, and officially also artist when it suits you.

JG: We like to call ourselves context hackers.

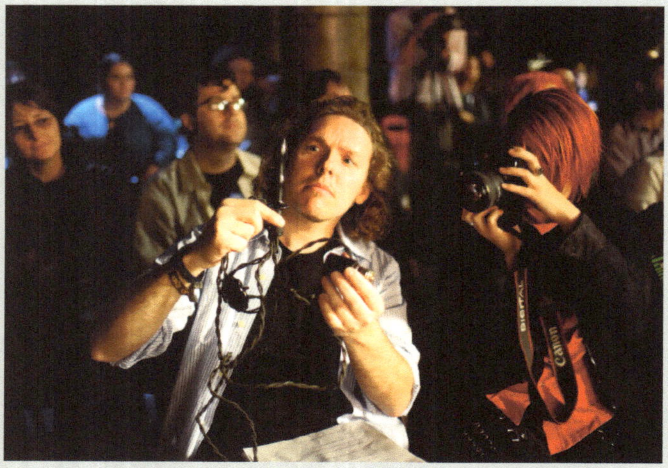
92. Johannes Grenzfurthner at Arse Elektronika 2007. Photo: Scott Beale (aka Laughing Squid)

RD: [laughs] I forgot to tell you that your talk at TEDxVienna[6] made me laugh so much. So much.

JG: Thanks a lot. But you kind of get the point. What we at monochrom are doing is trying to find the perfect medium for a certain message. We are a political group, so we try to spread information—political information, philosophy, technology, whatever it is—and for certain things you need a certain context, a certain medium. Sometimes what works best to distribute our work and ideas is a text file; sometimes it is a short film. Other times we can do a musical or a computer game, or we set up a prank in a public space. It depends. It's hard to nail us down with the medium we are using because what matters is our approach.

RD: And, of course, you also are the organizers of Arse Elektronika?[7]

JG: Oh yes [laughs].

RD: You know, I thought about Arse Elektronika this morning. I don't remember if it was in the TED talk or some other online interview with you I saw on YouTube, but you were saying that Arse Elektronika is about sex. It made me realize that we don't have enough sex projects in our book. In fact, we have zero. Are there any sex projects or sex-related projects you think we could or should include in the sprint book?

JG: There are many, there are many. Especially if you are talking about hacking. Arse Elektronika is about sex and technology, and the interaction between the two. And that brings you back to the printing of the first Bible, because Gutenberg financed some Bible printing by printing erotica at the same time. So there is always a correlation between new forms of media and pornography. VHS is a good example for that, but also the first Polaroid camera, which was called the Swinger[8] because they were targeting the amateur porn markets in the 1960s.

Broadband Internet would not exist the way it is now without porn consumers, especially male customers. The early days of the Internet were very male-driven. There is always a relationship between sex and technology. What is really interesting at the moment is the whole area of fucking machines and sex technologies, dildos and all that stuff. It's quite interesting to see that women have a pretty good..., let's say, connection to sex technology, because almost all

6. http://www.youtube.com/watch?v=K2Rvh8VG3o8
7. http://www.monochrom.at/arse-elektronika/
8. http://en.wikipedia.org/wiki/Polaroid_Swinger

the women I know have dildos, but men don't. Even more interestingly, I did a couple of interviews in Arse Elektronika with a couple of women, and some of them feel more interested in having sex with a robot than in having sex with a man.

RD: Oh nooo.

JG: Yeah, really. Because they have control over that. It's an interesting aspect of third wave feminism.

RD: They just don't know my boyfriend.

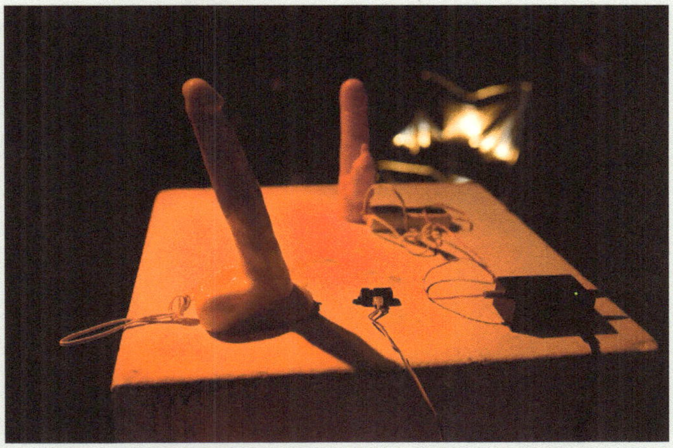

93. Arse Elektronika 2008. Photo: Scott Beale (aka Laughing Squid)

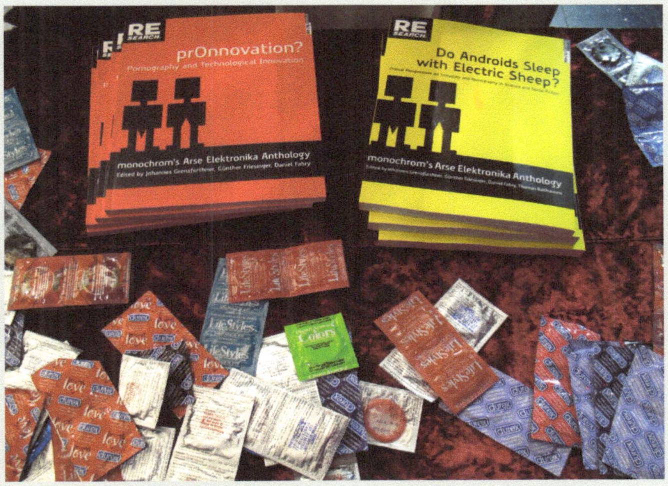

94. Arse Elektronika 2009. Photo: Scott Beale (aka Laughing Squid)

JG: In the do-it-yourself and hacker spaces scene, more and more women are interested in building their own sex toys, and playing with that, than there are men willing to build sex toys. For example, a good friend of mine, Kyle Machulis, is actually the project I would like to recommend. Kyle Machulis[9] is from Oklahoma, but now he lives and works in San Francisco, and he is one of the main guys of do-it-yourself sex technology. In his day job, he is working for a company that is creating a self-driving car for the DARPA challenge. That's his daytime job. Before that, he was working for Linden Lab—on Second Life's sex DIY interfaces, for example. Right now he is looking into cheap vibrators, cheap dildos, and the various interfaces available. He gets them, rips them apart, and builds completely new sex toys out of them. He shares his stuff online with people. What he is doing is pretty much like a new field of teledildonics and open source sex technology and DIY sex technology. You really should look into Kyle Machulis and his projects. His website is called slashdong.[10]

RD: Oh! I know slashdong.

RD: Now a question to satisfy my personal curiosity: Did someone from Ars Electronica in Linz ever react to Arse Elektronika?

JG: No.

RD: No?!?

JG: Interestingly, some of them invited us last year, so we gave a talk at Ars Electronica about Arse Elektronika.

RD: [laughs]

JG: We did that last year. It was actually pretty successful. It started off as a joke, because we were thinking that in its thirty years of existence, Ars Electronica had only once tried to do something about sex and the future; they called it "Next Sex,"[11] but it was very disappointing. They even had some crazy biologist there talking bullshit about rape as a natural strategy. So we decided that we had to do a conference on the subject ourselves. We wanted to bring together people who never meet, such as the people in hacker spaces who tinker

9. http://www.nonpolynomial.com/about/
10. http://www.slashdong.org/

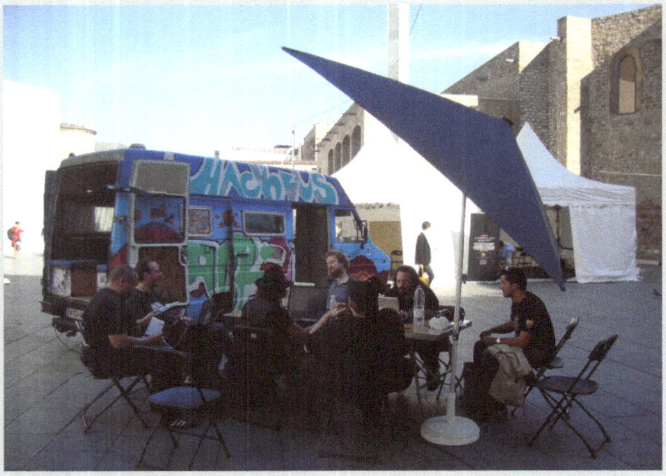

95. Hackbus. Photo: Stephanie Doll

around with sex toys and academics who specialize in gender theory, and have them discuss together. Arse Elektronika is designed to be like a meeting-place where crazy people from all over the planet can talk about sex and technology.

RD: How do you see the role of monochrom when it comes to technology. Do you see monochrom as a divulgator? A facilitator? A critique?

JG: I think we are definitely using technology because we grew up with that and that's what we use. For a certain period of time, people could call themselves digital artists just because they were using e-mail, but, of course, that is not possible anymore. The main problem that people have—and I'm not talking about politicians and stuff like that—is that some people also in the classical art world have these strange ideas that the Internet is yet another medium that's out there. But it's not another medium: it's another place of living, as I would call it. It's not like there was radio, then TV, and now there's the Internet. No, with the Internet there is an awfully dramatic interpretation, but I use it. It's where we are living, and as long as you can accept that there are people using it as their living quarters to some extent, you will fail to interpret that. In the meantime, we have players like Anonymous out there, who grew out of this interesting "fortune" community. It is so hard to explain to people what fortune means, and how fortune works, and how out of fortune Anonymous can emerge. It takes five to ten minutes to explain what it all means. But politicians are judging all of that. We have people like the American vice president who say there are digital terrorists. It is obviously clear that no one ever explained to him what is going on here. They [politicians] have a completely different view and [have chosen] wrong vectors to analyze here. That's the real cultural divide that's happening here. It's not the cultural divide of being online or not being online, but about how you are interested in what's going on online and how you actually want to learn about what's going on. And that's what I was talking about in my TED talk, that the world is getting more and more complex. You really have to understand that kind of stuff to really be able to create interest in creative activists. If you do not understand that, you fail [on an epic scale].

RD: I just saw that monochrom is touring in a Hackbus?!!??

JG: One of my biggest critiques of the hacker scene and the hacker spaces is that they are exclusionist white boys' clubs. It is interesting to see that in the digital art world there are many women doing interesting stuff, but most of them do not know or do not go to hacker conferences or hacker spaces, because different mechanisms operate in the hacker circle and in the digital art circle. And they don't like each other most of the time... but it's bullshit. They have to talk and they should talk.

One of our current projects is called the Hackbus,[12] a mobile hacker space. Static hacker spaces usually end up in white male clubs because there are certain psychological and sociological factors that keep women, minorities, or simply other cultures out. For example, there is a really great hacker space in Germany, but they don't have a single Turkish member, yet the space is located in the Turkish neighborhood, because it's cheap to rent there. Of course, class and sex barriers exist in pretty much all the subcultural scenes you can imagine. But the ideology of the hackers space is "do it yourself"—if you

11. *Next Sex — Sex in the Age of its Procreative Superfluousness* http://90.146.8.18/en/archives/festival_archive/festival_overview.asp?iPresentationYearFrom=2000
12. http://hackbus.info/index.php/Main_Page
13. http://metalab.at/wiki/English

want to do stuff, you show up and we help you—and they define their space as an "open space." But to [create] an open, inclusionist space, you have to be really active. People don't just show up magically. You have to actively invite them, have a reach-out program, and most of the hacker spaces don't have that. Hackbuses take the interesting idea of hacking, DIY culture, and digital art, and bring it where the public is. In Vienna, for example, we have a hackers space called MetaLab,[13] and we did a Hackbus tour last year where we went to the twenty-three districts in Vienna. We choose one public space, park the Hackbus there, and suddenly all these people are there building and playing around with Arduino, doing interesting stuff. These people would never ever go to the hackers space in Vienna because they don't know of it. If they did know about it and went, they would probably be stared at and leave immediately. That's why we like the idea that you can go to where the people actually are. It's an experiment in educational hacking. We are [going] beyond the educational system that was formed for the purpose of creating workers for the Industrial Revolution.

RD: That's a great project. Thank you so so much for your time, Johannes.

JG: No problem, no problem. Hope to see you again soon.

RD: Me tooooo.

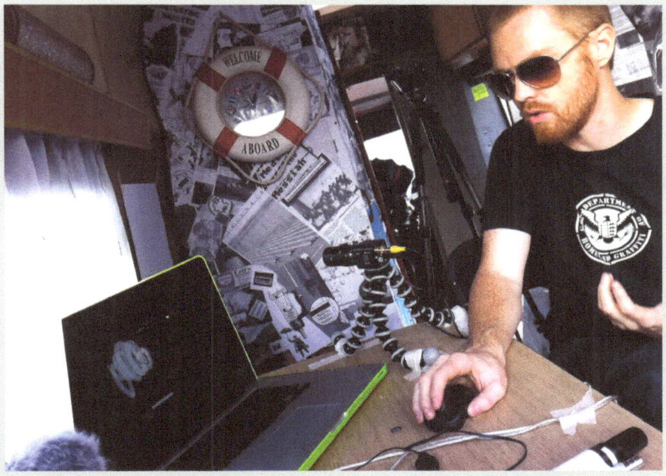

96. Evan Roth inside the Hackbus. Photo: Johannes Grenzfurthner

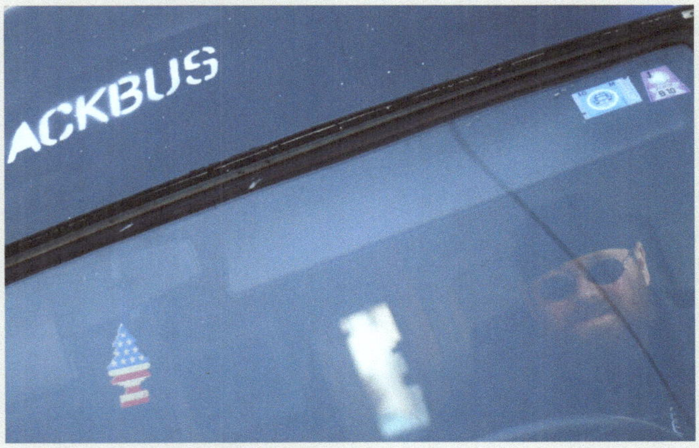

97. Hackbus driver. Photo: Samuel Huron

SEBASTIAN BRAJKOVIC
LATHE CHAIRS IV
2008

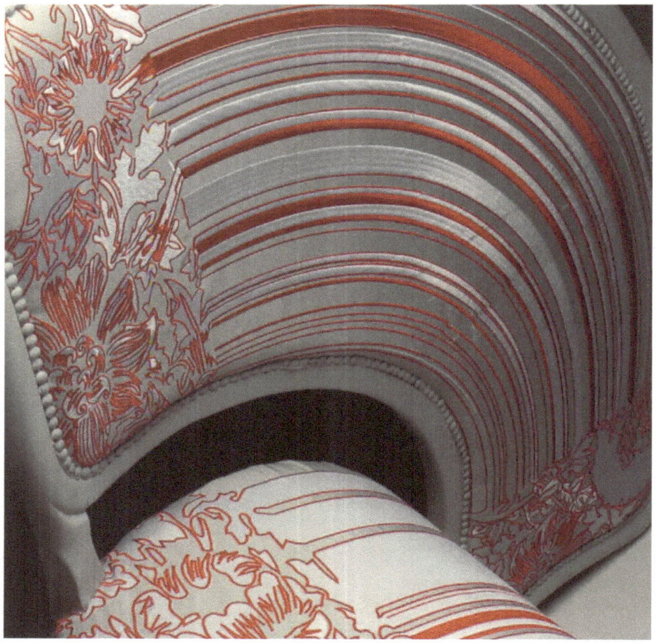

26. Sebastian Brajkovic, *Lathe Chair IV*, 2008

Starting with nineteenth-century chair remnants, Sebastian Brajkovic marries these familiar designs with new digitally created forms. His chairs appear stretched and contorted, preserving the periphery of the originals as bookends of the new object.

The "chairs" are not chairs, but expressions of operations that with a turn or twist introduce the algorithmic into antique forms. The new objects are not made of wood, as the referent furniture would imply, but rather cast in bronze. The additional solidity of the work—the unexpected excessive mass—not only turns the chair into a sculptural object, but also inverts the ephemeral digital operation it depicts.

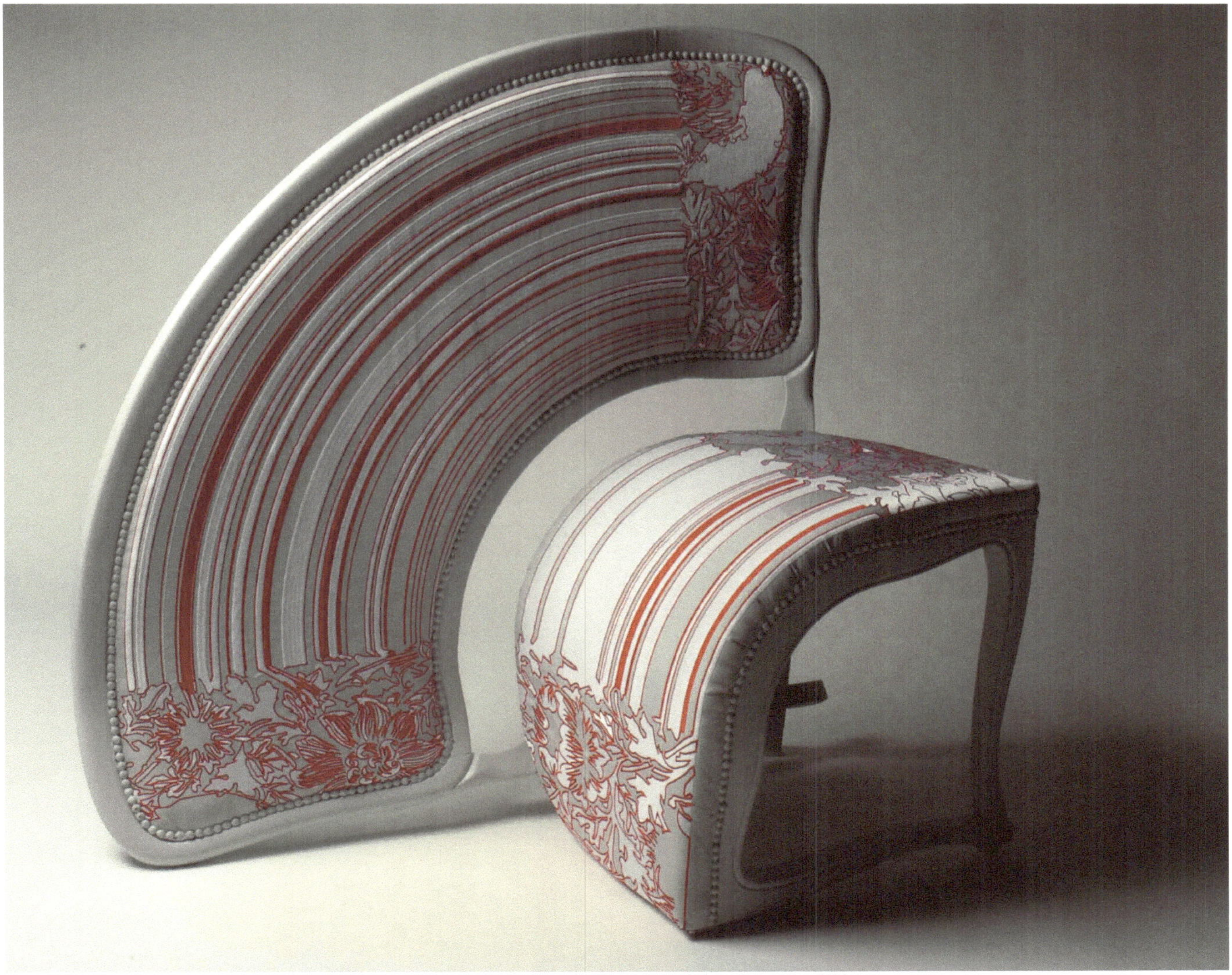

27. Sebastian Brajkovic, *Lathe Chair IV*, 2008

JULIUS VON BISMARCK
IMAGE FULGURATOR
2007-2008

"The Image Fulgurator is a device for physically manipulating photographs. It intervenes when a photo is being taken, without the photographer being able to detect anything. The manipulation is only visible on the photo afterwards."[1]

Hidden from the photographer's gaze, the wielder of the Image Fulgurator lies in wait for its prey: a camera flash. As a digital camera snaps an image, the Image Fulgurator, a "weaponized" camera, inserts an image into the shot. By using a simple flash detector, the Image Fulgurator can flash a millisecond-long image into an unsuspecting camera's field of view.

The "weapon," built by Berlin-based artist Julius von Bismarck, is really a camera in reverse. Rather than take in light through the lens and direct it onto a film plane or digital array, it projects images through the lens and out to the world. But unlike a projected image, the cast image comes and goes as fast as a camera shutter, perfectly synchronized with the technological eye but invisible to human vision.

The Image Fulgurator is best used to corrupt the tourist experience, as masses of sightseers gather and robotically photograph landmarks and "places of interest," often with loved ones in the photos, as souvenirs. Perched among the tourists, resembling a normal camera on a tripod, the Image Fulgurator quietly subverts sightseeing once tourists look in their digital camera display to find an unexpected message.

1. http://www.juliusvonbismarck.com/fulgurator/idee.html

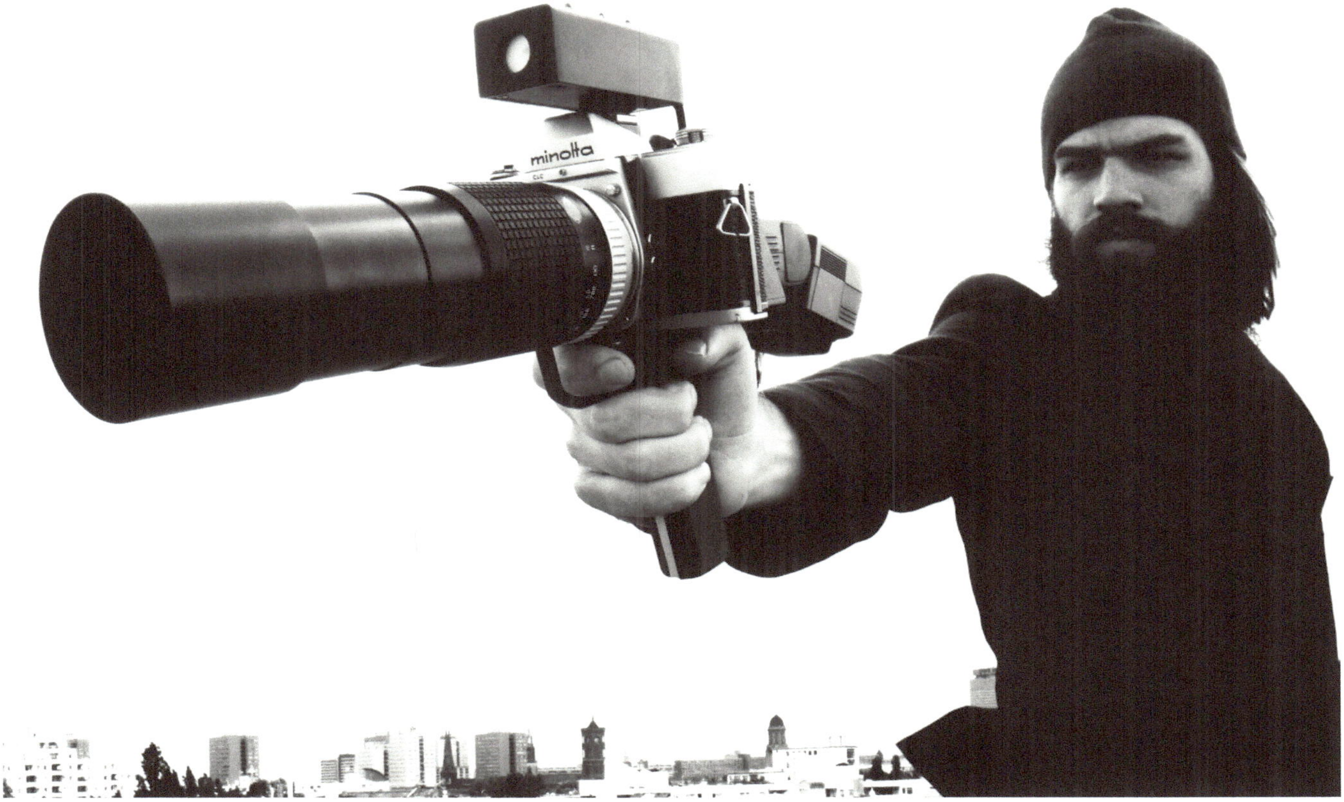

28. Julius von Bismarck with the Image Fulgurator, 2007-8

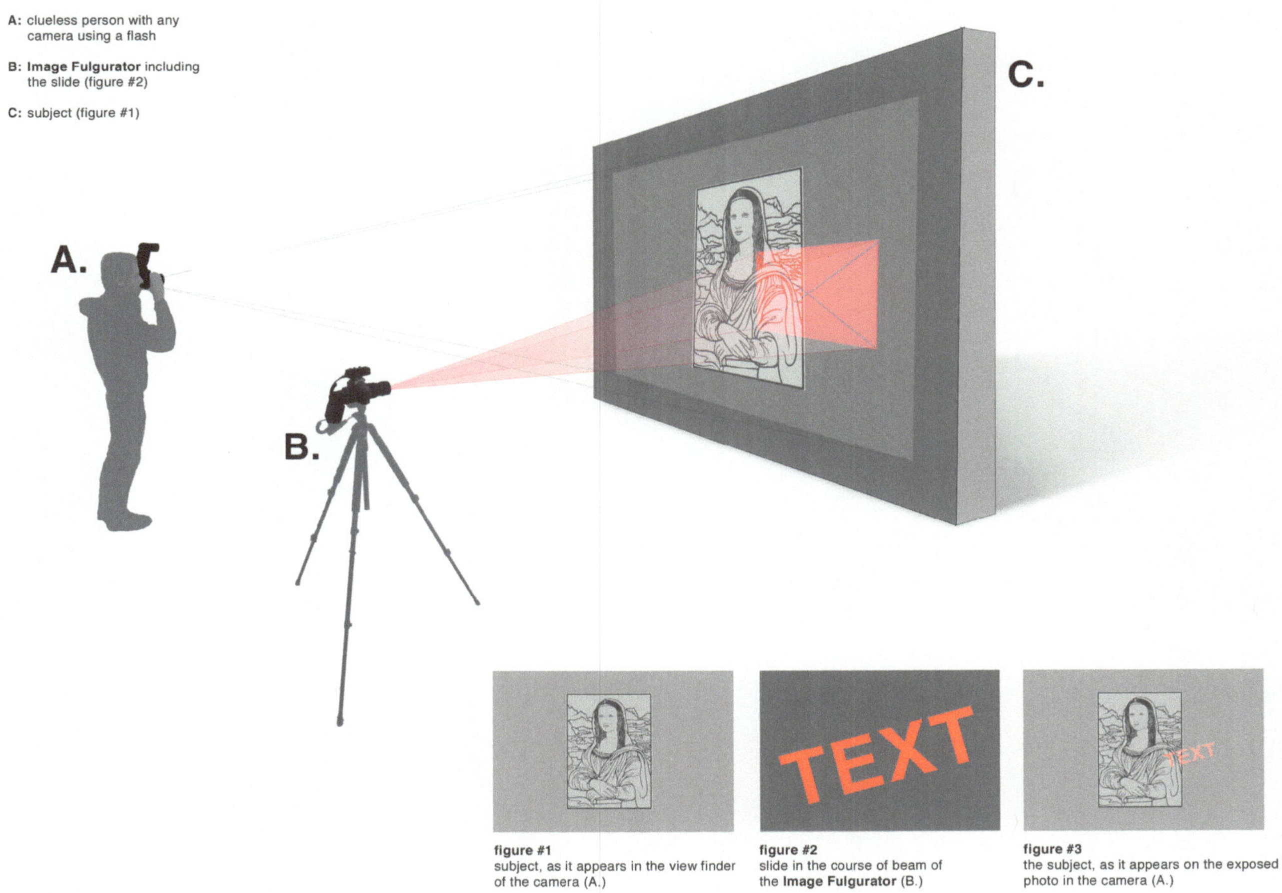

29. A diagram showing the Image Fulgurator's function

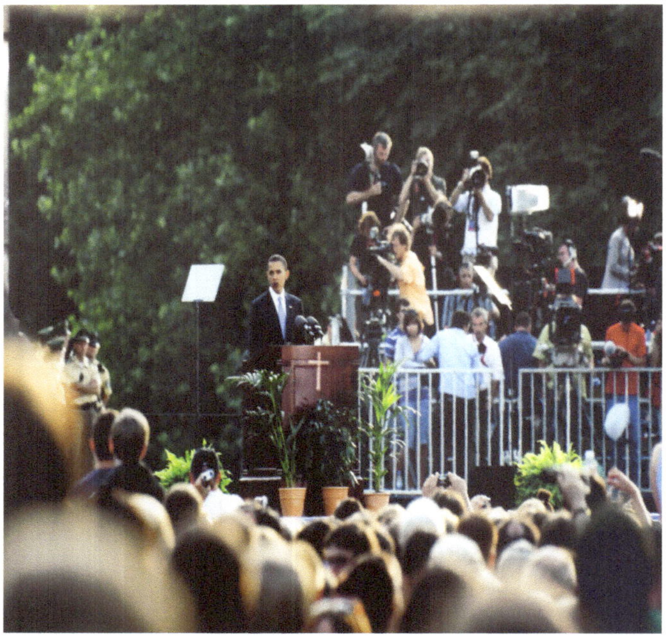

30. Barack Obama's speech in Germany in front of Berlin's Siegessäule, July 24, 2008

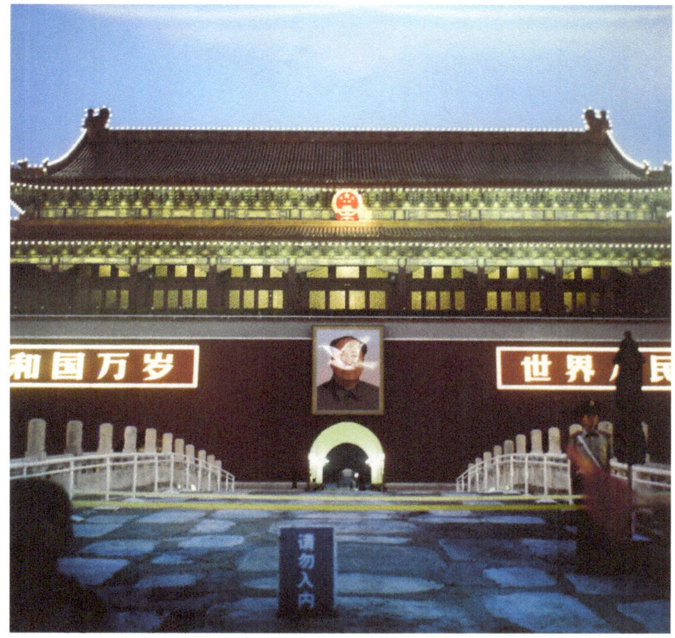

31. Fulguration of "the Magritte dove" on the Mao Zedong portrait at Tiananmen Square, Beijing

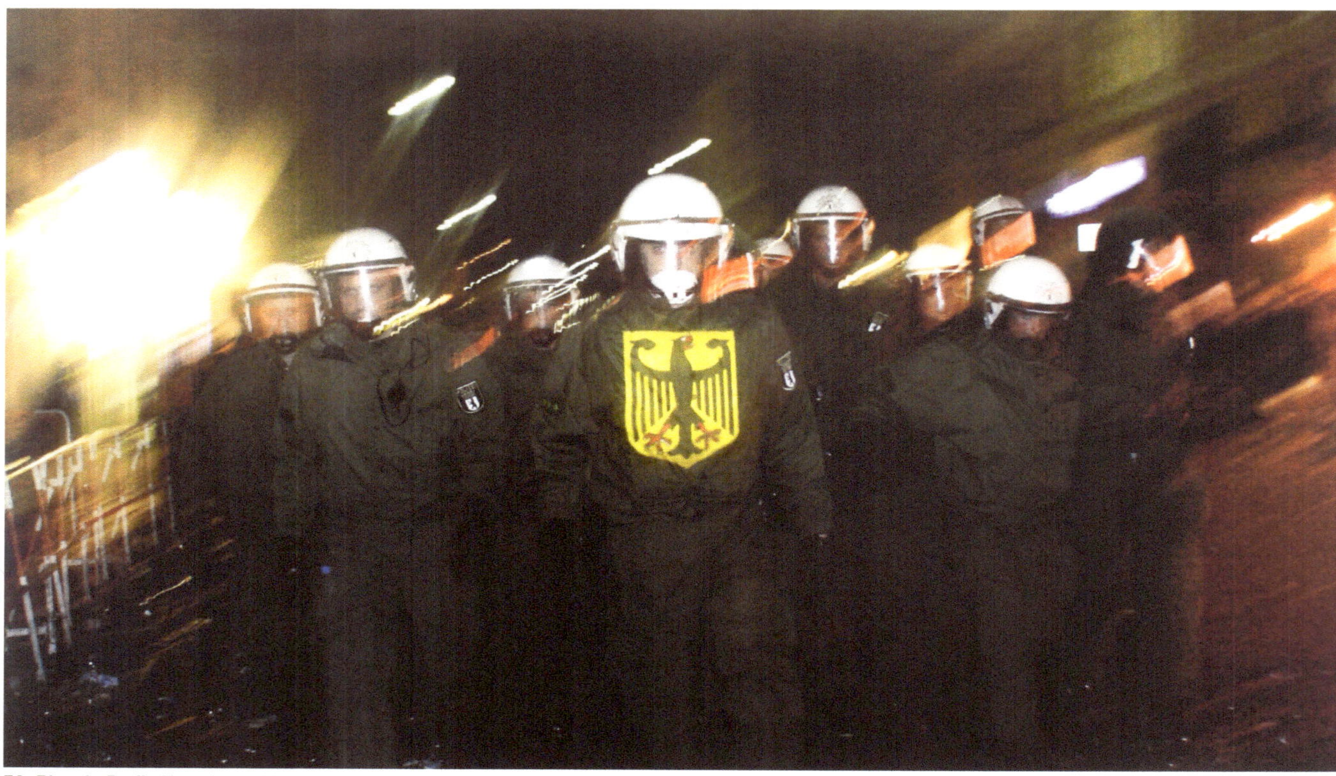

32. Riots in Berlin Kreuzberg, May 1, 2009

The Conscience of a Hacker

by
+++The Mentor+++
Written on January 8, 1986

Another one got caught today, it's all over the papers. "Teenager arrested in Computer Crime Scandal", "Hacker Arrested after Bank Tampering"... Damn kids. They're all alike.

But did you, in your three-piece psychology and 1950's technobrain, ever take a look behind the eyes of the hacker? Did you ever wonder what made him tick, what forces shaped him, what may have molded him? I am a hacker, enter my world... Mine is a world that begins with school... I'm smarter than most of the other kids, this crap they teach us bores me... Damn underachiever. They're all alike.

I'm in junior high or high school. I've listened to teachers explain for the fifteenth time how to reduce a fraction. I understand it. "No, Ms. Smith, I didn't show my work. I did it in my head..." Damn kid. Probably copied it. They're all alike.

I made a discovery today. I found a computer. Wait a second, this is cool. It does what I want it to. If it makes a mistake, it's because I screwed it up. Not because it doesn't like me... Or feels threatened by me... Or thinks I'm a smart ass... Or doesn't like teaching and shouldn't be here... Damn kid. All he does is play games. They're all alike.

And then it happened... a door opened to a world... rushing through the phone line like heroin through an addict's veins, an electronic pulse is sent out, a refuge from the day-to-day incompetencies is sought... a board is found. "This is it... this is where I belong..." I know everyone here... even if I've never met them, never talked to them, may never hear from them again... I know you all... Damn kid. Tying up the phone line again. They're all alike...

You bet your ass we're all alike... we've been spoon-fed baby food at school when we hungered for steak... the bits of meat that you did let slip through were pre-chewed and tasteless. We've been dominated by sadists, or ignored by the apathetic. The few that had something to teach found us willing pupils, but those few are like drops of water in the desert.

This is our world now... the world of the electron and the switch, the beauty of the baud. We make use of a service already existing without paying for what could be dirt-cheap if it wasn't run by profiteering gluttons, and you call us criminals. We explore... and you call us criminals. We seek after knowledge... and you call us criminals. We exist without skin color, without nationality, without religious bias... and you call us criminals. You build atomic bombs, you wage wars, you murder, cheat, and lie to us and try to make us believe it's for our own good, yet we're the criminals.

Yes, I am a criminal. My crime is that of curiosity. My crime is that of judging people by what they say and think, not what they look like. My crime is that of outsmarting you, something that you will never forgive me for.

I am a hacker, and this is my manifesto. You may stop this individual, but you can't stop us all... after all, we're all alike.

+++The Mentor+++

33. Hacker's Manifesto, 1986

PAUL VANOUSE
SUSPECT INVERSION CENTER
2011

Paul Vanouse's *The Suspect Inversion Center* was an operational laboratory where the artist replicated the genetic (DNA) fingerprints of well-known subjects. Together with his assistant, Kerry Sheehan, the biomedia artist set up a temporary lab for the exhibition "Fingerprints..." held in the Project Space of the Ernst Schering Foundation in Berlin in 2011. Using Vanouse's own DNA as well as equipment anyone can buy on the Internet, the duo recreated in public performances the "genetic fingerprints" of O.J. Simpson as well as master copies of historical courtroom images of the athlete's DNA from his 1995 murder trial.

While the reliability of ballistic, bite-mark, and even fingerprint analysis can sometimes be questioned in courtrooms, genetic evidence is still widely regarded as the forensic gold standard. *The Suspect Inversion Center* points out that even DNA evidence can be fudged. "The Trial of the Century" with O.J. Simpson was not only the most publicized criminal trial in American history, but also the first time that a defense team had enough scientific understanding to successfully contest the validity of DNA evidence.

More recently, scientists in Israel have demonstrated that DNA evidence can be fabricated. "You can just engineer a crime scene,"[1] explained Dan Frumkin, lead author of a paper published by *Forensic Science International* in 2009. "Any biology undergraduate could perform this."

The Suspect Inversion Center invited visitors to witness the whole process of genetic forgery. Interestingly, this particular falsification differs from the forgeries that have plagued the history of art: in genetic forgery, the process and results negate the very authority of the original model.

1. http://www.nytimes.com/2009/08/18/science/18dna.html

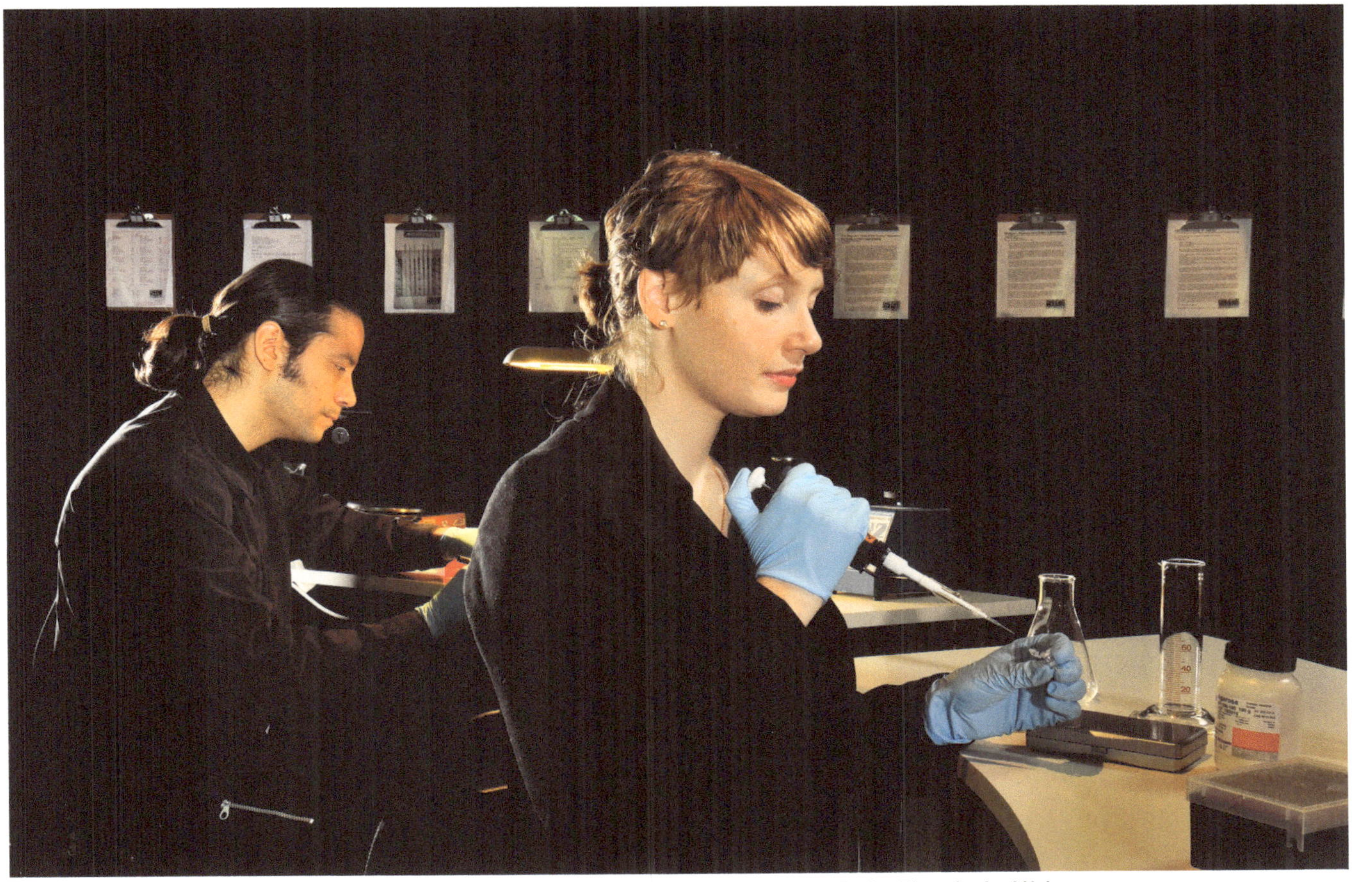

34. View of the exhibition Fingerprints... in the Project Space of the Ernst Schering Foundation, Berlin. Photo by Axel Heise

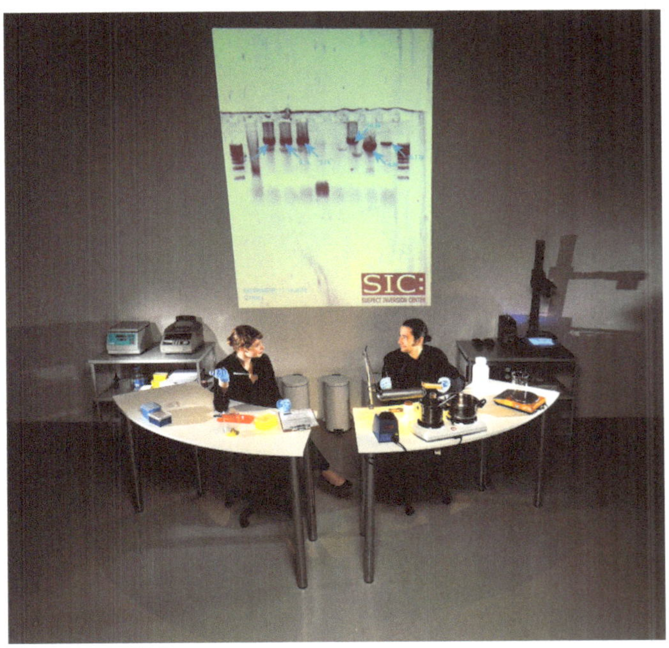

35. View of the exhibition Fingerprints... in the Project Space of the Ernst Schering Foundation, Berlin. Photo by Axel Heise

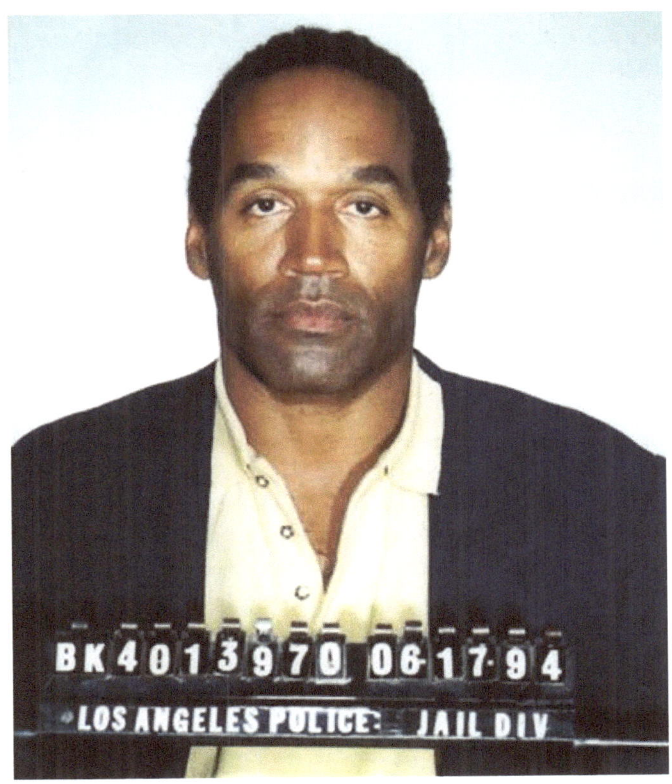

36. O.J. Simpson mugshot, 1994

JULIAN OLIVER AND DANJA VASILIEV
NEWSTWEEK
2011

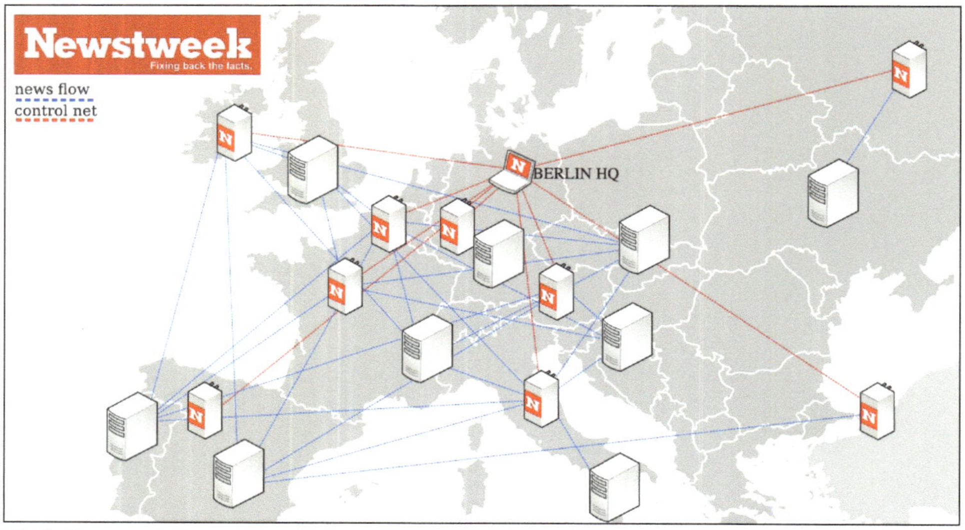

37. The Newstweek 'Control Net' disrupts the legitimate news flow network through nearly invisible WiFi hotspot devices. Julian Oliver and Danja Vasiliev, *Newstweek*, 2011

Picture the scene. You're sitting in your favourite café, enjoying an organic carrot juice and free WiFi. You open the BBC News or CNN website. But the headlines you're reading are bizarre, surreal, and in some cases downright shocking. Why would the US appoint Julian Assange as its head of Defense? Why would Bill Gates be sent to China to convince the country to invest in the US army?

This online mayhem can be attributed to the presence in your café of the *Newstweek* device, an object so insignificant looking that, as one of its developers puts it, "it either goes unnoticed or appears as part of the infrastructure. It's a bastard in beige." The device remotely manipulates news stories read by people browsing the web and using the same wireless hotspot. Once the unassuming *Newstweek* box is plugged into the wall, data can be slyly injected into the communication between the wireless router and a user. Design, logo, advertisements—everything appears as it should on the news websites. Everything but the headlines and content.

Newstweek, whose tagline is "don't read what you believe," investigates what artists Julian Oliver and Danja Vasiliev call "Network Insecurity" by approaching the network as a medium for rigorous, creative investigation. *Newstweek* points to the impossibility of true objectivity. As soon as a story hits the media, it is submitted to subtle and sometimes unconscious manipulations by the people who write it. The work also questions the domesticated forms of dissent provided by Facebook and blogs. Even when authors appear to bear their biases with pride, these relatively new platforms often preach their ideas and opinions to the converted. *Newstweek* asks instead, how far can we push the limits of subjectivity? Why not allow dissent to reach new territories?

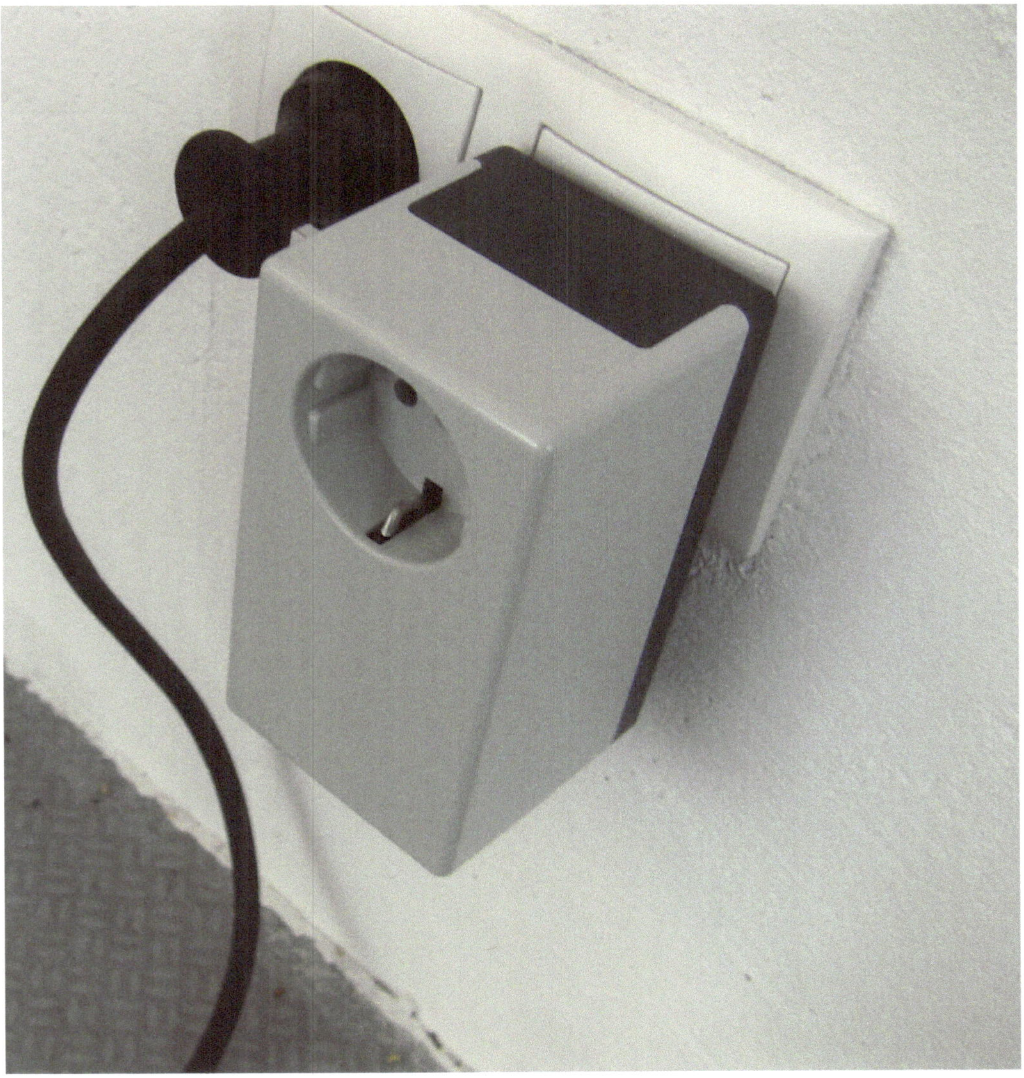

38. The Newstweek device discreetly plugs into any standard outlet, hiding in plain sight

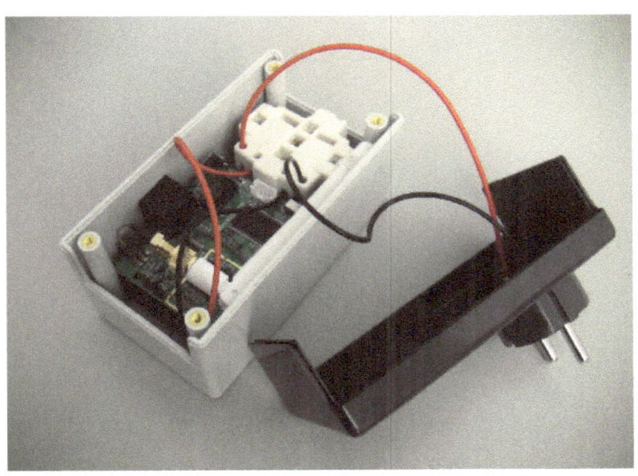

39. Hardware hidden inside the Newstweek device

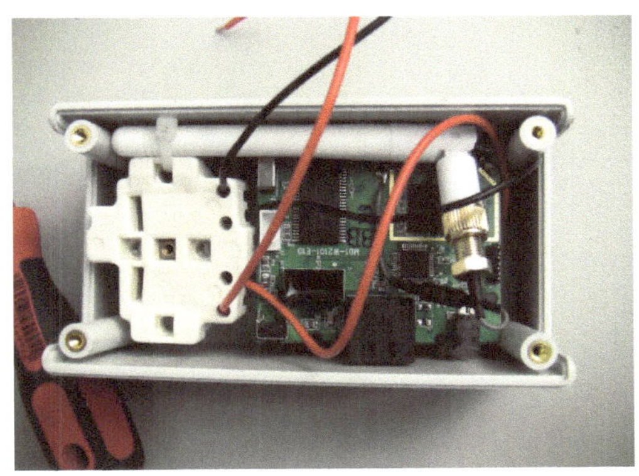

40.

38–40. Julian Oliver and Danja Vasiliev, ***Newstweek***, 2011

Julian Oliver and Danja Vasiliev

As an artist, how do you see your role in a technological or scientific setting?

We believe the most transformative language of our time is that of engineering, one that shapes the way we communicate, move, earn and even think.

As Critical Engineers, we take the technological ready-mades and assumptions of today and apply radical 'patches.' These patches pervert, reposition, rework and in some cases even improve what are otherwise accepted as immutable technological givens, engineered by industry with our best interests in mind.

MARCO DONNARUMMA
GOLDEN SHIELD MUSIC
2010-2011

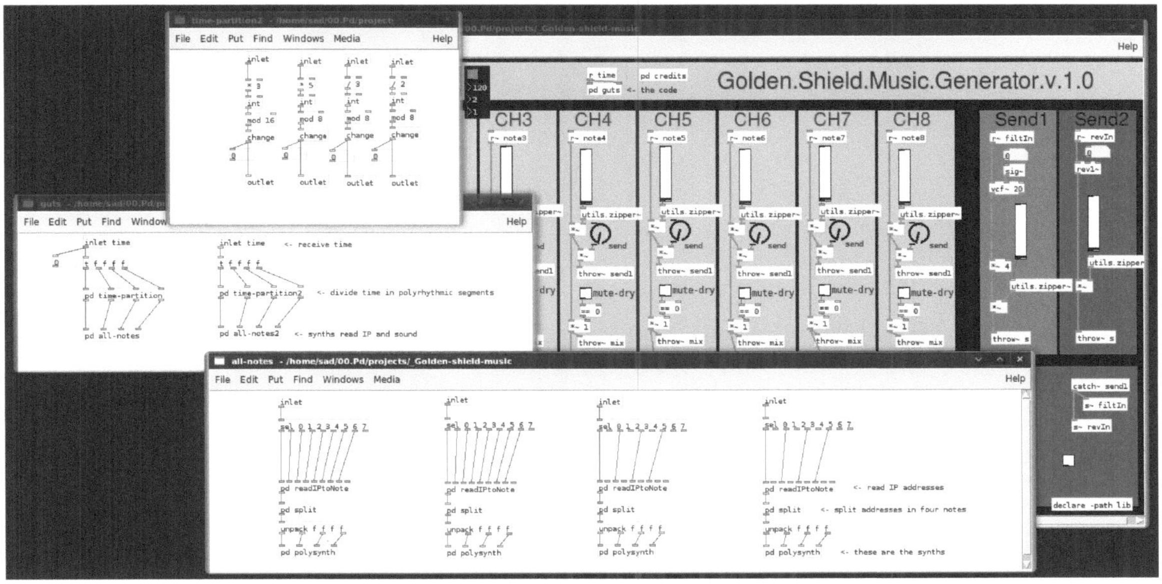

41. Golden Shield Music Generator v.1.0, GUI and guts, Marco Donnarumma, 2009

Marco Donnarumma's *Golden Shield Music* uses censorship technology to generate a sound composition. The objective of the project is not to produce remarkable compositions but to raise awareness about the phenomenon of web censorship.

Golden Shield Music collects the twelve website Internet Protocol (IP) addresses that are most often blacklisted by the Chinese government, processes them through an automated MIDI polyphonic synthesizer, and generates music for playback in eight audio channels. The project establishes an abstract relationship between Internet information and musical algorithms.

Whereas the name *Golden Shield Music* refers to China's Golden Shield Project,[1] aka the "Great Firewall of China," Donnarumma's project also alludes to Internet censorship that is more widespread than we might be aware of. Both Australia[2] and the UK[3] are notorious for their governments' attempts at censoring online content. In Italy, Silvio Berlusconi's government has been trying[4] to block[5] The Pirate Bay as a "preventive measure." In Finland, activist Matti Nikki's website lapsiporno.info monitors[6] his country's censorship program. As many read on the front pages of numerous international newspapers only a short while ago, Egypt attempted to block Twitter as protesters were taking to the streets. Even more recently, news has emerged that the regime in Iran has increased censorship and online disruption to block anti-government protests. The list goes on and on, but it is even more worrying that Internet monitoring apparently could even go global. During a presentation he made at Piksel,[7] the eighth festival for Electronic Art and Technological Freedom in Bergen, Norway, Donnarumma pointed to Recorded Future,[8] funded[9] by the CIA and Google, as a disquieting project that monitors tens of thousands of websites, blogs, and Twitter accounts in real time to find relationships between people, organizations, actions, and incidents. The goal of this intense data mining is to "predict the future" by "looking at the 'invisible links' between documents that talk about the same, or related, entities and events." Imagine what could happen if the CIA were to detect "invisible links" between you and an organization it suspects of terrorism.

1. http://en.wikipedia.org/wiki/Golden_Shield_Project.
2. http://en.wikipedia.org/wiki/Internet_censorship_in_Australia.
3. http://en.wikipedia.org/wiki/Internet_Censorship_in_the_United_Kingdom.
4. Italy blocks access to The Pirate Bay, http://www.geek.com/articles/law/italy-blocks-access-to-the-pirate-bay-20080811/.
5. Italy Blocks The Pirate Bay Yet Again, http://www.techdirt.com/articles/20100207/2246518070.shtml.
6. http://en.wikipedia.org/wiki/Lapsiporno.info.
7. http://www.piksel.no/.
8. https://www.recordedfuture.com/.
9. Exclusive: Google, CIA Invest in 'Future' of Web Monitoring, http://www.wired.com/dangerroom/2010/07/exclusive-google-cia/.

WILLY SENGEWALD (THEGREENEYL)
JAMMER HORN
2008

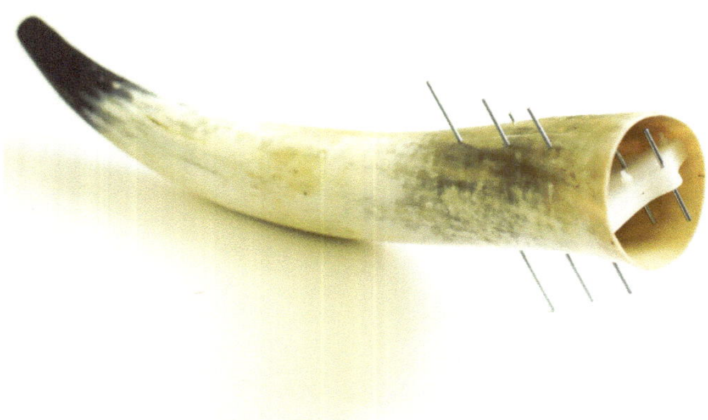

42.

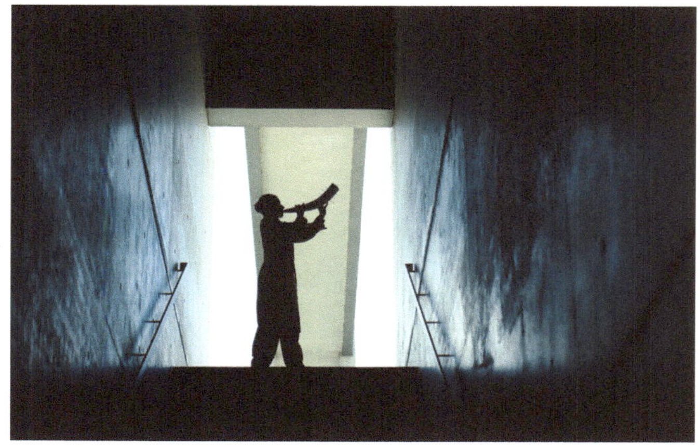

43.

42-43. Willy Sengewald (TheGreenEyl), *Jammer Horn*, 2008

Willy Sengewald, a member of TheGreenEyl based in Berlin and New York, fitted a horn with a high-range cell phone jammer that is triggered by a temperature sensor.

Blowing into the horn grants you the immediate power to suppress any cell phone activity taking place within thirty to fifty meters. *Jammer Horn* turns upside-down the function of an instrument used in ancient times to establish communication between humans and to mediate relationships with the gods. Moreover, this onetime tool of communication offers an opportunity for revenge: it denies modern communication devices the right to operate.

The horn, integral to certain mythologies and religions, thus becomes a signal-bearer for a new generation's interaction with its environment.

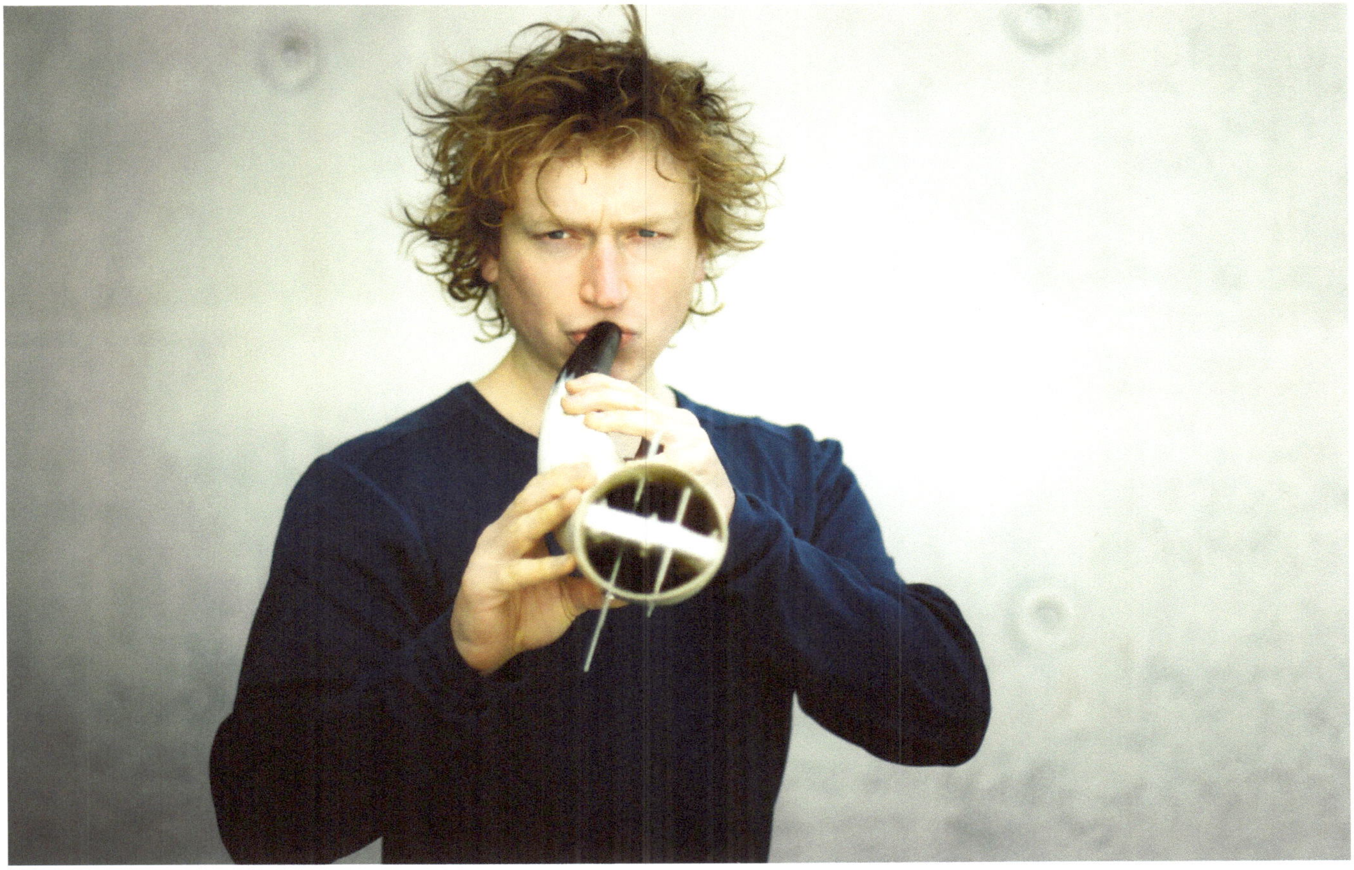

44.

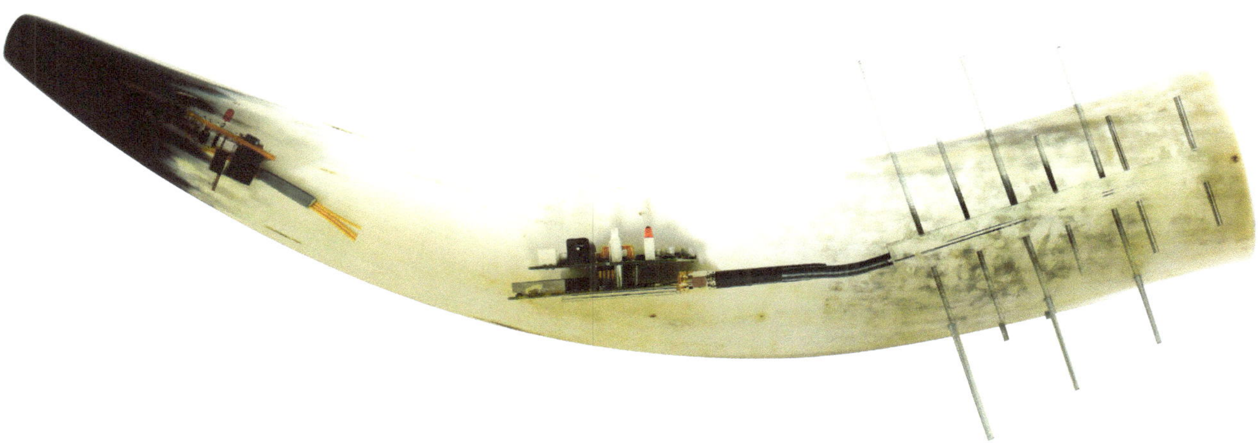

45.
44–45. Willy Sengewald (TheGreenEyl), *Jammer Horn*, 2008

BOREDOMRESEARCH
REALSNAILMAIL
2008

RealSnailMail is a project by British duo Vicky Isley and Paul Smith, otherwise known as boredomresearch, which uses real snails with RFID chips glued to their shells to carry and deliver electronic messages at their own "snail's pace."

A message sent through the *RealSnailMail* website will travel at the usual speed to a server, where it is entered into a queue. It will wait there until a living snail crawls into proximity with one of the RFID readers that acts as a "dispatch center." The reader identifies the snail from its chip, checks its availability, and assigns the message at the top of the queue. The references to the electronic message are then physically carried around the tank by the snail until it passes close to a second reader. If ever this happens, the second reader identifies the message and forwards it to the recipient's email address.

At the time of this writing, the average time for a message to be delivered via *RealSnailMail* is 572 days, twelve hours, seven minutes, and twenty-one seconds. For context, consider that letters delivered in the 1800s by the Pony Express took an average of ten days to arrive. *RealSnailMail* is not just "slow;" it represents something close to geological time in our age of rapidly networked communications.

There are many ways to read this work—a farcical take on the "slow" movement, or a comment on our information-addled cultural climate and need for immediacy—but at its core, it's something as absurd as it is possible: hacking a snail.

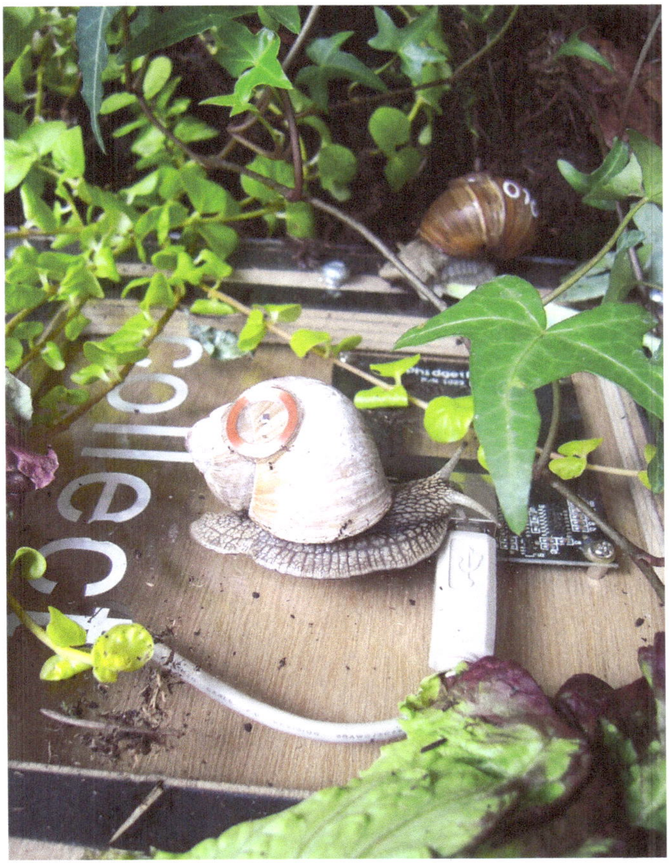

46.

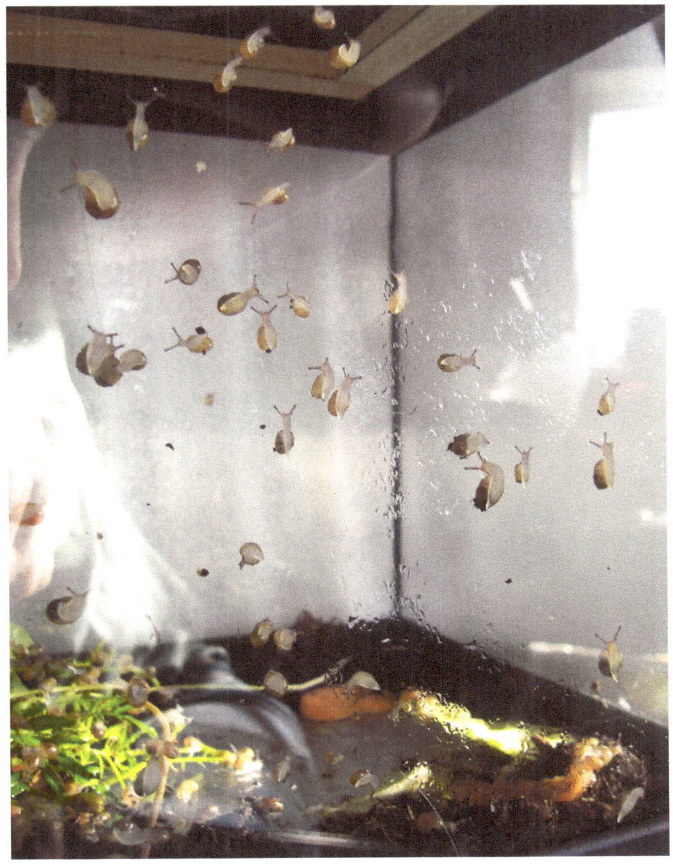

47.

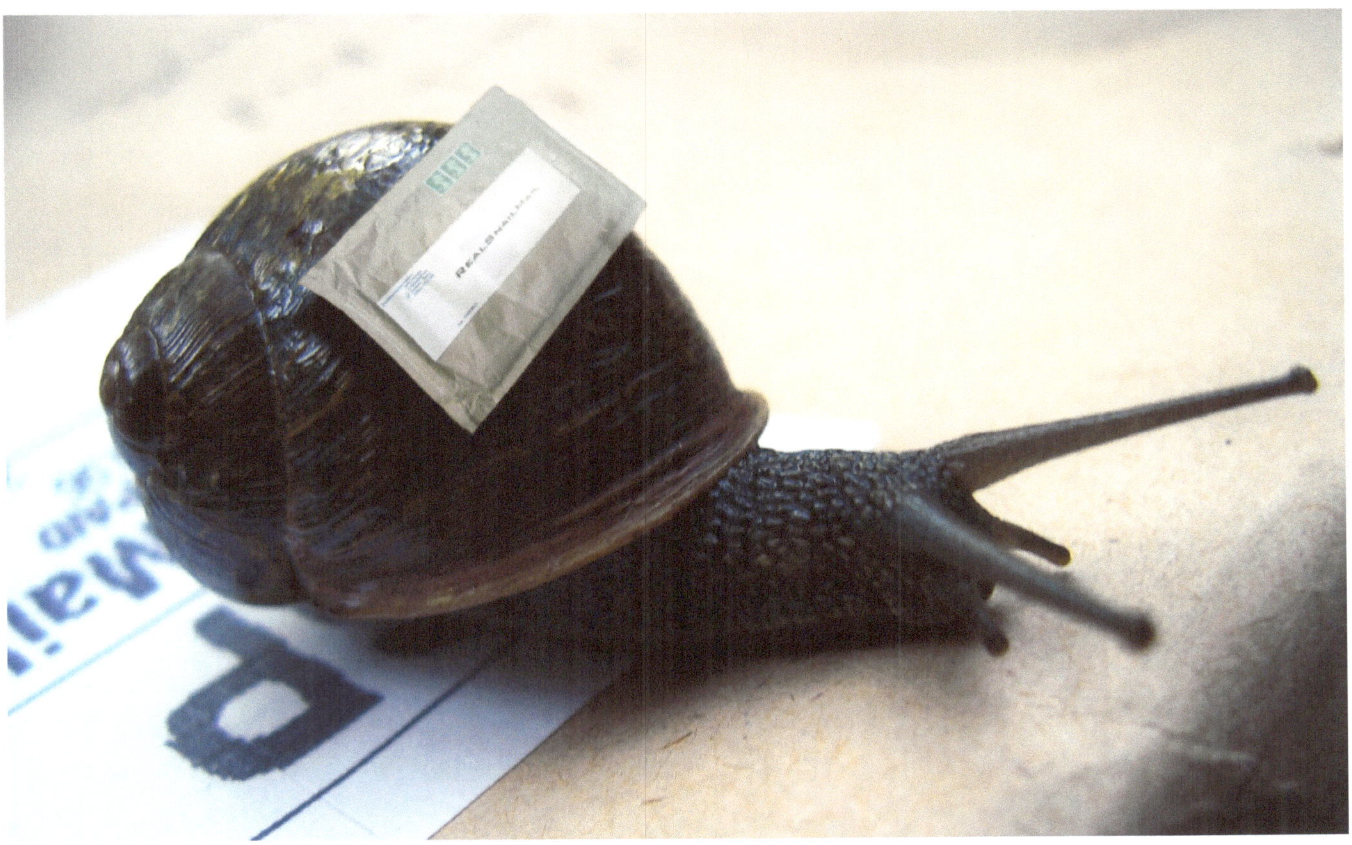

48.

46–48. boredomresearch, *RealSnailMail*, 2008

hacker: n.

originally, someone who makes furniture with an axe

1. A person who enjoys exploring the details of programmable systems and how to stretch their capabilities, as opposed to most users, who prefer to learn only the minimum necessary. RFC1392, the Internet Users' Glossary, usefully amplifies this as: A person who delights in having an intimate understanding of the internal workings of a system, computers and computer networks in particular.

2. One who programs enthusiastically (even obsessively) or who enjoys programming rather than just theorizing about programming.

3. A person capable of appreciating hack value.

4. A person who is good at programming quickly.

5. An expert at a particular program, or one who frequently does work using it or on it; as in 'a Unix hacker'. (Definitions 1 through 5 are correlated, and people who fit them congregate.)

6. An expert or enthusiast of any kind. One might be an astronomy hacker, for example.

7. One who enjoys the intellectual challenge of creatively overcoming or circumventing limitations.

8. [deprecated] A malicious meddler who tries to discover sensitive information by poking around. Hence password hacker, network hacker. The correct term for this sense is cracker.

The term 'hacker' also tends to connote membership in the global community defined by the net (see *the network*. For discussion of some of the basics of this culture, see the How To Become A Hacker FAQ. It also implies that the person described is seen to subscribe to some version of the hacker ethic (see *hacker ethic*).

It is better to be described as a hacker by others than to describe oneself that way. Hackers consider themselves something of an elite (a meritocracy based on ability), though one to which new members are gladly welcome. There is thus a certain ego satisfaction to be had in identifying yourself as a hacker (but if you claim to be one and are not, you'll quickly be labeled bogus). See also *geek, wannabee.*

This term seems to have been first adopted as a badge in the 1960s by the hacker culture surrounding TMRC and the MIT AI Lab. We have a report that it was used in a sense close to this entry's by teenage radio hams and electronics tinkerers in the mid-1950s

Def: Hacker, Jargon File, Version 4.4.8, www.catb.org/jargon

CITIZEN SCIENCE

AMATEUR, AUTODIDACTIC, DEMOCRATIZATION OF SCIENCE, HOME LABORATORIES, GARAGE SCIENCE, CROWDSOURCING, DISTRIBUTED COMPUTING

"Contemporary science has its roots in the achievements of amateur scientists of centuries past. Although they lacked what we would define as formal scientific training, they deciphered the basic laws of physics and principles of chemistry. They invented instruments. And they discovered, documented, sketched, and painted planets, comets, fossils, and species."
—"Amateur Science, Strong Tradition, Bright Future,"
by Forrest Mims III[1]

Citizen science[2] is a term used to describe scientific work completed by individuals who may not have specific scientific training, but are engaging in the work regardless through their own heuristic process. Projects are sometimes issue oriented and facilitated by technology, online networks of volunteers, or open source collaborators with the goal of contributing to larger problem solving, data collection, mapping, or public awareness. In other instances, the term citizen science is used to describe self-initiated, unconventional, hands-on studies. Still other examples of citizen science may be oriented toward education and playful experimentation. In all of these cases, this methodology calls for work outside of traditional research environments.

While its connection to the discipline of contemporary art may be recent, citizen science is not new. Many long-running citizen science projects have had a meaningful impact

1. *Science* 284, no. 5411 (2 April 1999): 55-56; available at http://www.sciencemag.org/content/284/5411/55.full.
2. *Science* will be used here as a shorthand for "science and technology."

on the world of research and institutional science; one of the longest-running such projects is the Audubon Society's "Christmas Bird Count," a yearly census of birds in the western hemisphere conducted almost entirely by volunteer birders. Such projects, while often supported by large institutions like research universities, government agencies, and nonprofit institutes, rely on public participation to further scientific research goals.

Citizen science projects range from myriad bird-watching undertakings instigated by Cornell University to distributed computing projects like SETI@Home, Stardust@Home, and Galaxy Zoo run by the SETI Institute, NASA, and independent agencies—all of which entreat the public to contribute to scientific research. Some citizen science projects operate in a liminal, boundary-dissolving manner themselves: witness the popular online game "Foldit," which uses the pretext of a computer game to empower the public to identify possible ways a chain of amino acids folds into a natural three-dimensional shape, a notoriously difficult determination. Top-ranked Foldit players excel at identifying possibilities, often better than their computer counterparts.

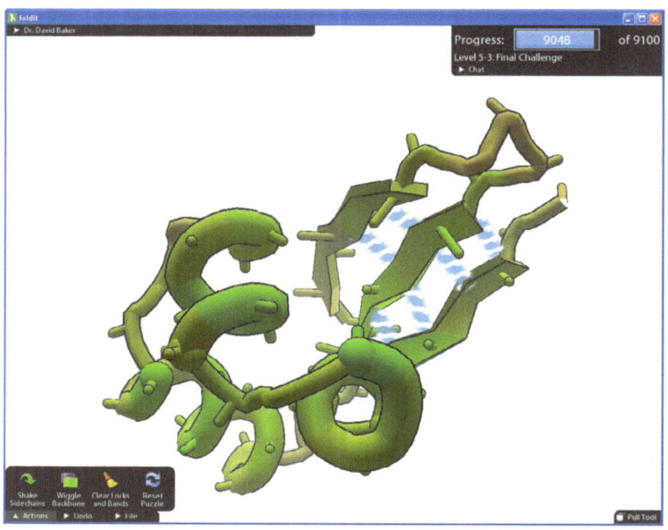

49. Screenshot of the experimental protein-folding game, Foldit

The more that academic and research institutions engage with the public in this way, the more porous the division between the "two cultures"[3] becomes, allowing for far greater disciplinary freedom of movement. If we view artists working at the intersections of art, science, and technology as having a certain responsibility to demystify science and empower the public to both question and participate in it, then the jump from this kind of "crowdsourced" citizen

3. Physicist and novelist C.P. Snow in his 1959 Rede lecture at Cambridge University, "The Two Cultures," provocatively argued that the breakdown in communication between the sciences and the humanities should be remedied or it would remain a major hindrance to solving the world's problems. His subsequent publications, *The Two Cultures and the Scientific Revolution* (Cambridge: Cambridge University Press, 1960) and *The Two Cultures: and a Second Look* (Cambridge: Cambridge University Press, 1964), have become canonical texts in discussions of interdisciplinarity.

science to the outright adoption of scientific practices in an art context is not a large one.

Like the citizen science projects mentioned above, the artists and micro-institutions in this chapter all engage the public, whether through a direct call to action, a museum-like presentation of materials, or an emphasis on public connection through open events and workshops. This is the place of dirty hands and excited conversations, strange confluences of subjects, and evenings spent peering into homemade microscopes.

Just as institutional citizen science projects ask the public to lend a hand—making people feel as though their input might well affect how research is conducted—artists practicing as citizen scientists engage the public in a new inclusive relationship. The effect is empowering. Visitors to Machine Project in Los Angeles may come for the art and leave with knowledge of marine biology or processing; participants in CRITTER salon's "Enormous Microscopic Evening" enter an art museum and suddenly find themselves identifying insects or understanding the operating principles of microscopy.

They may sometimes be more slapdash than scientists, but artists implementing this methodology can identify research paths and points of engagement that might be ignored by scientists. And as they are less beholden to funding and utility, they also need not adhere to the cultural narrative of peer-reviewed experimental rigor. As instigators of citizen science, artists can raise questions that are not more or less significant than those posed by their white-coated counterparts—but that are certainly strange, unorthodox, and often compelling.

CESAR HARADA
PROTEI
2010-ONGOING

Protei is a fleet of pollution-collecting sailing drones, developed primarily to collect oil spills. It was designed by artist Cesar Harada and an international team of contributors under Open Hardware licensing by Open_Sailing, randomwalks, V2, and Amorphica. The inflatable drones are intended to be inexpensive, semiautonomous, self-righting, hurricane-ready alternatives to current oil spill collection technology, which, according to Harada, collects only three percent of the spilled material.

Being an open source project, the sailing drones are designed to be adapted and reimagined within different environmental contexts. While *Protei* gathers oil spilled into the open sea, other versions could tackle the "plastic island" in the North Pacific gyre or toxic substances in urban waterways. *Protei*—and its parent project, Open_Sailing, an international organization attempting to design and build a bio-architectural "International Ocean Station"—emphasizes the involvement of multitudes of people to achieve things that, in other contexts, might cost millions of taxpayer dollars and produce semi-secret information for closed groups of scientists. "It is not a utopian project," Harada writes. "We are working on it every day."

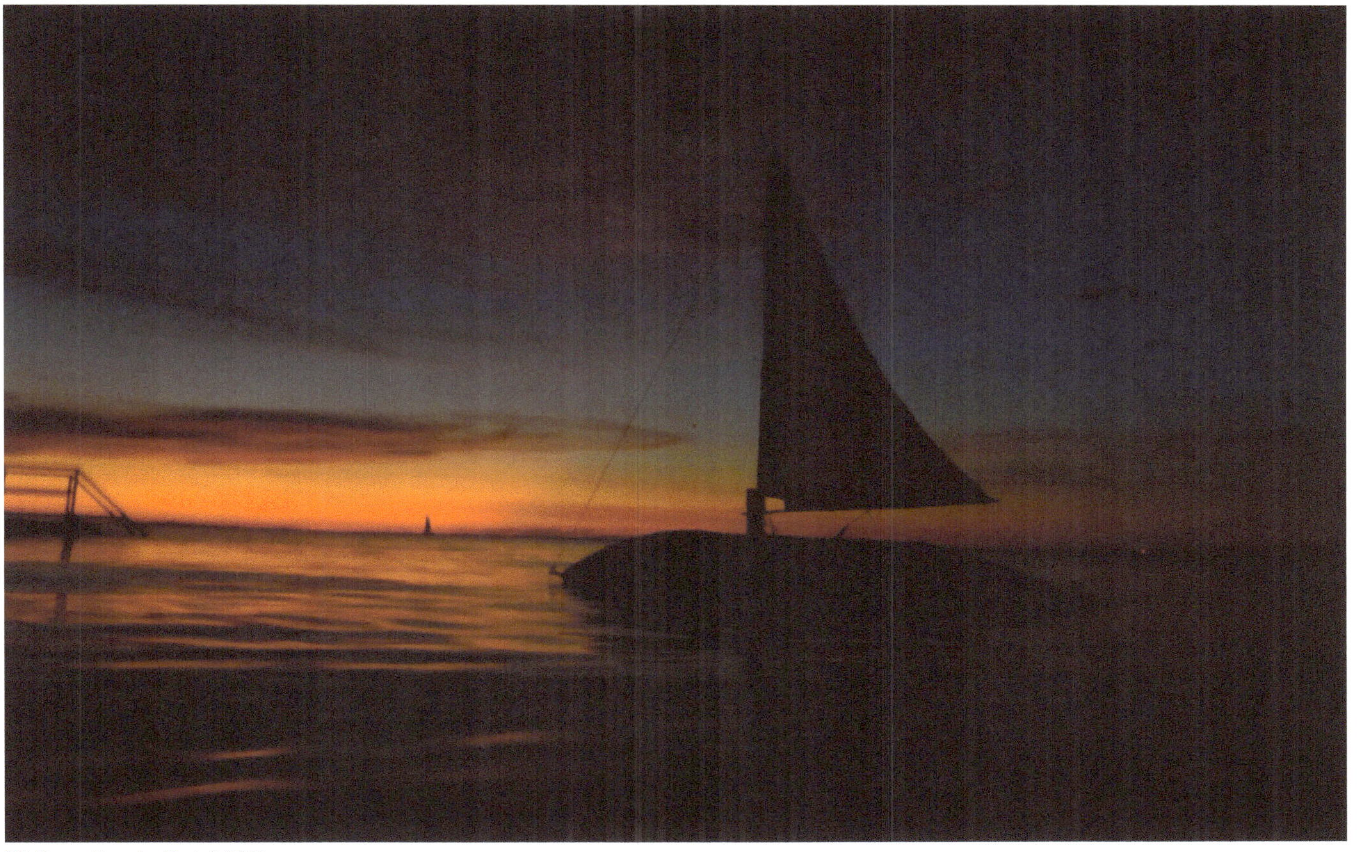

50. Cesar Harada, *Protei*, 2010

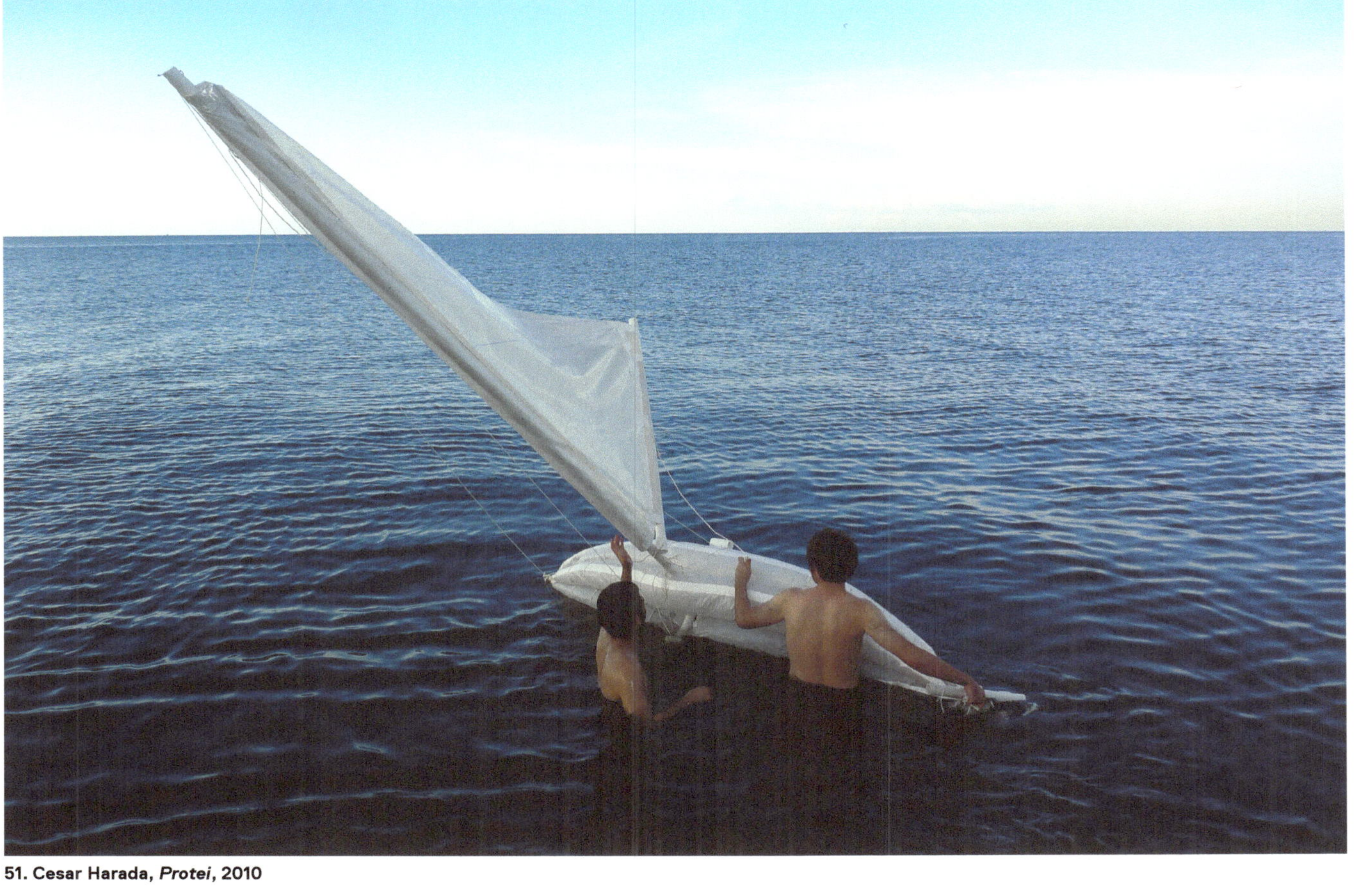

51. Cesar Harada, *Protei*, 2010

Cesar Harada

As an artist, how do you see your role in a technological or scientific setting?

My answer is personal and reflects only my perspective as an artist in a technological or scientific environment. Before I get more specific, it is necessary to set this question in the context of accelerating technological developments at the beginning of the 21st century, when science resembles and often overpasses our wildest fantasies. I have abandoned consuming fictions since I became interested in science, and I now live in a world that is not limited by technology, but that is powered by technology, without boundaries. The future is open as technology and science reconfigure the environment, the larger ensemble that they influence, the Anthropocene.

Often, the general public mistakingly thinks of science and art as oppositional; one would be logic and procedural— the other intuitive and anarchist; one would be precise—the other scrappy. Such pre-conceptions are inaccurate and counterproductive. The history of art and the history of science are undistinguishable.

I won't elaborate about the historical evolution of the art of engineering, the science of emotions, or demonstrate that both art and science utilize extremely creative investigations methods. I would rather be more practical and list a few roles an interchangeable artist/scientist can take in the development of technology/art research.

The making of technological or scientific research is a complex process that requires many components. An artist/scientist, as defined above, can act in two main areas: inside the science, and around the science.

Within the science

- **generates the technology/science, tinkering.**
- **participates in the research process as collaborator, host/resident.**
- **makes a critical and creative contribution.**
- **builds prototypes, researches devices, tests, makes mistakes, observes, writes papers, makes progress.**
- **elaborates on research strategy, explores alternative methods of investigations.**
- **manages, creates connections between ideas, principles, and people with different areas of specialization.**
- **develops applications for the technology, science principles.**
- **plugs into other technology, couples.**

Around the science

- **shares findings/ideas with the research community.**
- **contextualizes, historically, socially, and practically, how to implement the science into a technology.**
- **expands/narrows the scope of the research.**
- **makes commentary, fantasizes, makes scenari of technological hedonism/dystopia.**
- **hacks the technology, tests its boundaries, plays.**

— subverts, outreaches for science/technology.
— disturbs, trolls, lobbies.
— promotes, fundraises, writes grants.
— advises ethics board.

Unfortunately these days I spend more time "around the science" creating the conditions to research than I spend doing the actual research. I personally have much more interest in acting "within the science" and generating the science/technology. I may enroll in a research program again very soon, to become part of larger research group that allows me to focus on making science and technological progress, leaving the "around the science" work to other people who have interest in being in that space.

As a conclusion I would quote Joseph Beuys' dictum, "Everyone is an artist" and add "Everyone is a scientist." The boundaries between art and science are arbitrary; post-industrial society is the product of our technology and imagination. Beyond deconstructivism, we are in a time of competition and collaboration for personal and general progress. The question, "Is it art or is it science?" does not matter anymore, and the question, "Are you a scientist or are you an artist?" does not matter anymore either. Every artist is a scientist. Every scientist is an artist. We are all people.

What I think matters is that technology, science and art are all contained in the Environment, Nature, the larger ensemble. Anything that positively serves the environment is worth researching. As a person, either considered an artist or scientist, I just keep this in mind, and make it my role as a human.

HEHE (HELEN EVANS AND HEIKO HANSEN)
NUAGE VERT
2003-2009

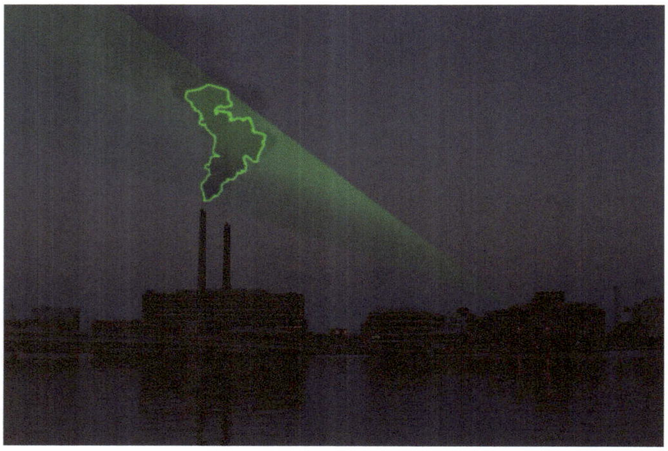

52. HeHe, *Nuage Vert*, 2007 ©HeHe

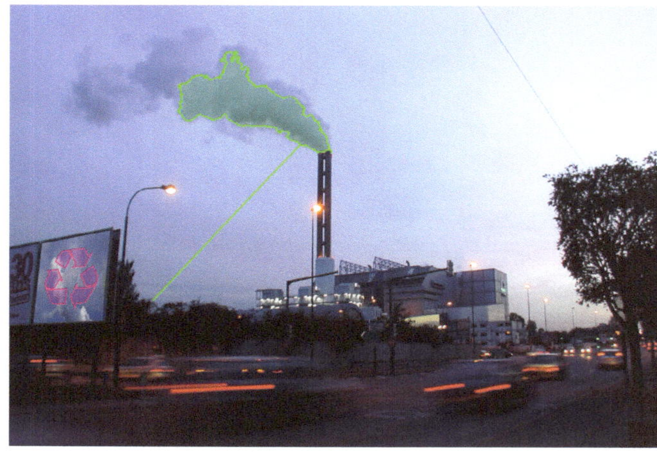

53. HeHe, *Nuage Vert*, 2004 ©HeHe

Created by HeHe duo Helen Evans and Heiko Hansen, *Nuage Vert* turns the vapor emissions of a household waste incinerator or an energy plant into an amorphous projection surface. The system uses a moving laser beam to draw the outline of the cloud issuing from the plant. Developed in collaboration with experts in laser technology, computer science, electrical engineering, energy production, and air quality monitoring, the work is a physical form of info-aesthetics.

By making large-scale pollution brightly visible, the project brings critical attention to its production. "Waste is relegated to the outskirts of the city, along with everything that we find inconvenient, and infrastructures are kept in the shadow as much as possible," explains HeHe to French newspaper *Libération*.[1] "Modern incinerators are conceived so that their emissions are not visible. This leads to a form of 'deresponsibilization.' Our project attempts to make more visible the chain of waste management. If we want to change perceptions and attitudes on a daily citizen level, we need aesthetic stimuli as much as we need educational messages."

HeHe projected *Nuage Vert* multiple times between 2003 and 2009. For each performance, the artists made an effort to involve the immediate community affected by the industrial pollution, including plant personnel, local authorities, environmental watchdogs, residents, cultural organizations, and schools.

1. Marie Lechner, "Le rayon vert divise Saint-Ouen," *Libération*, Feb. 6, 2009, http://www.liberation.fr/culture/0101570790-le-rayon-vert-divise-saint-ouen.

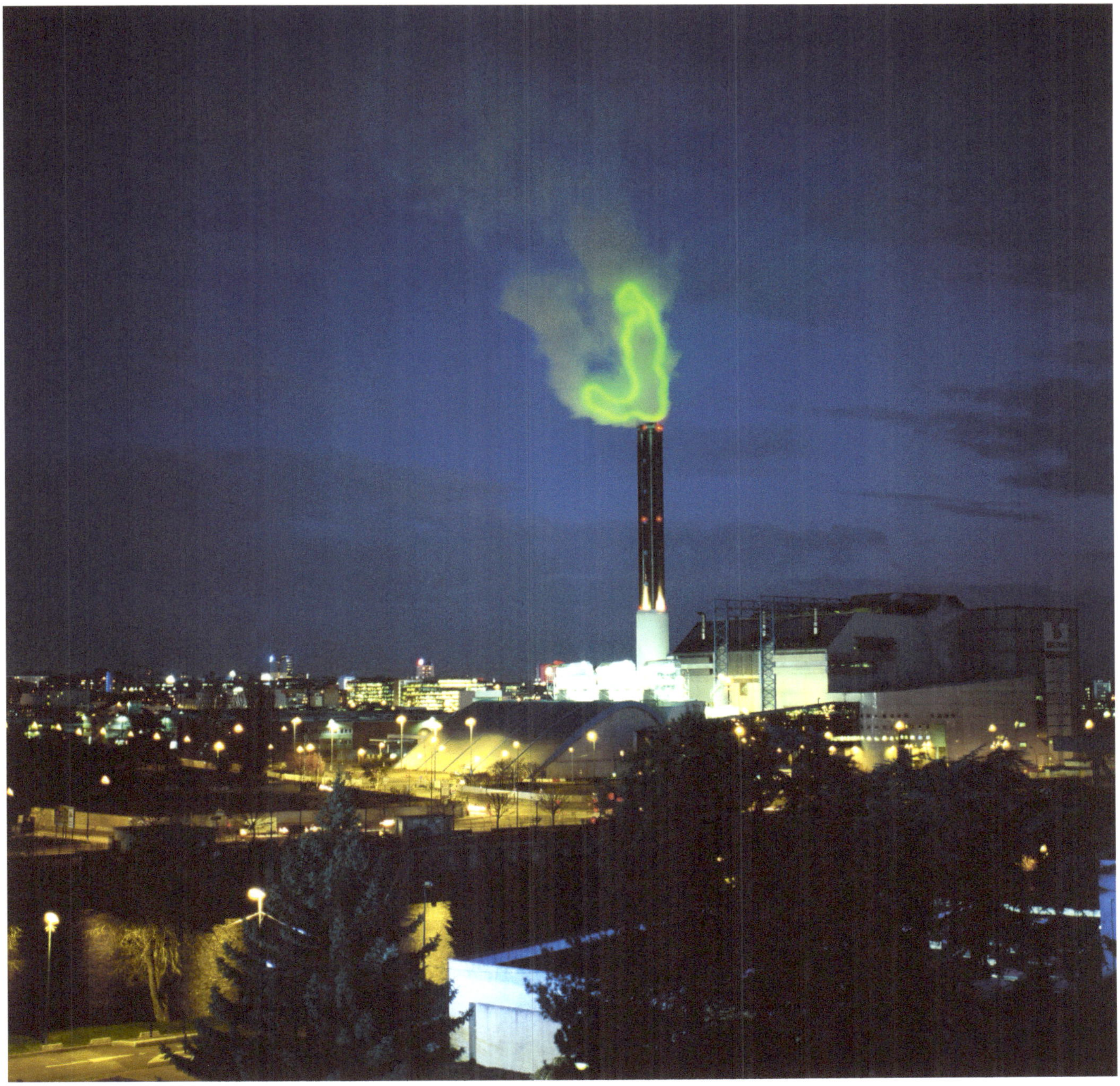

54. HeHe, *Nuage Vert*, 2009 ©HeHe

CRITTER
ENORMOUS MICROSCOPIC EVENING
2010

"Enormous Microscopic Evening" was a traveling event held at the San Francisco Exploratorium and the Hammer Museum in Los Angeles in 2009 and 2010, respectively. Organized by CRITTER Salon (with Machine Project), a San Francisco-based artist group dedicated to expanding the relationships between culture and the environment, the evening was all about microscopes: building them, manipulating them, and peering into the unseen recesses of the world through them. CRITTER called the event a "microscope jam session."

The microscopes involved in the event ranged from homemade devices and simple water lenses to state-of-the-art microscopes that truly allowed participants to see the invisible. Organisms, live cells, insects, plant matter, objects brought from home—very little was safe from magnification. With the assistance of various scientific and natural history organizations, the CRITTER Salon held workshops in DIY microscope construction and demonstrated how a new age of "social microscopy" is transforming the sciences.

The focus (so to speak) on microscopes presents them as a metaphor for what a scientific or technological approach can bring to the practice of art: if an artist's job is to look at things in an unexpected way, then how better to sharpen and intensify that gaze than to use a microscope? To see the unseen, to reveal hidden truths about the world—this is the work of both scientist and artist.

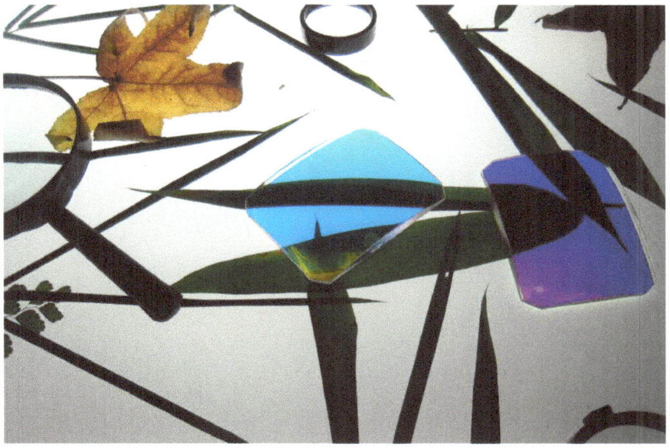

55.

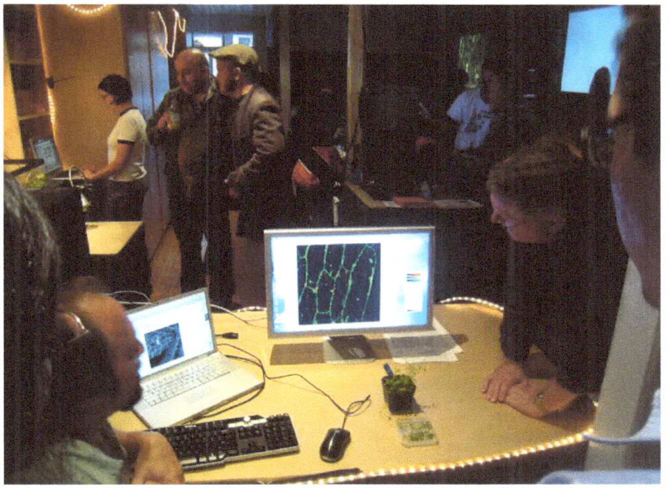

58.

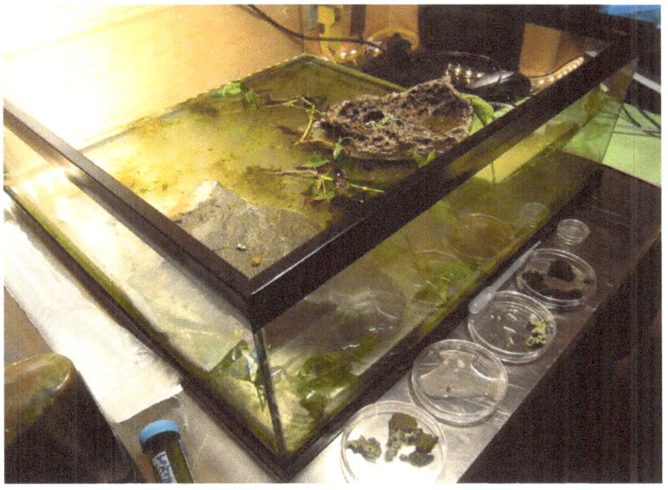

56.

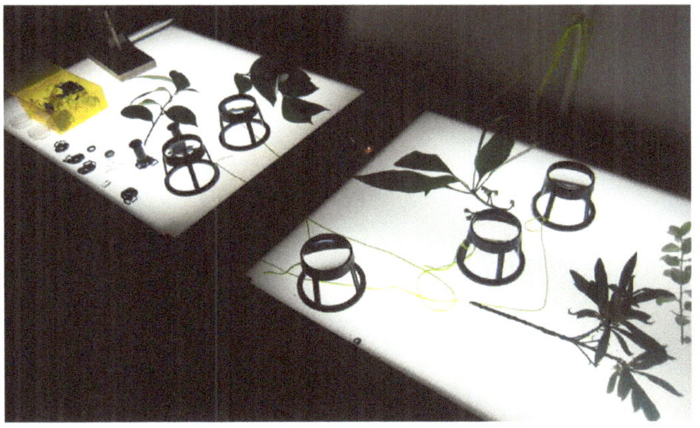

59.

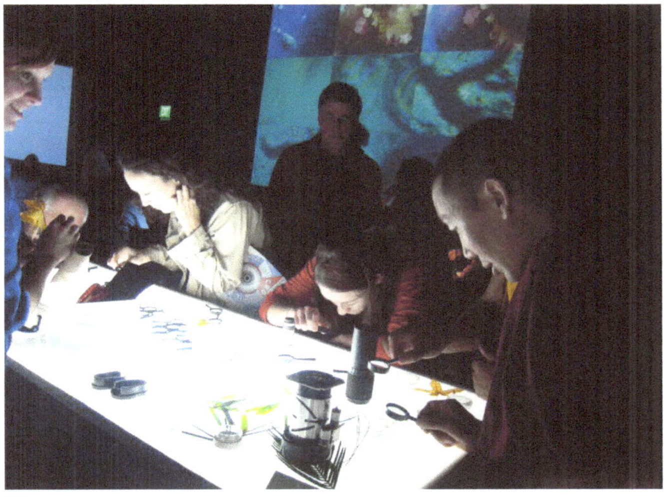

57.

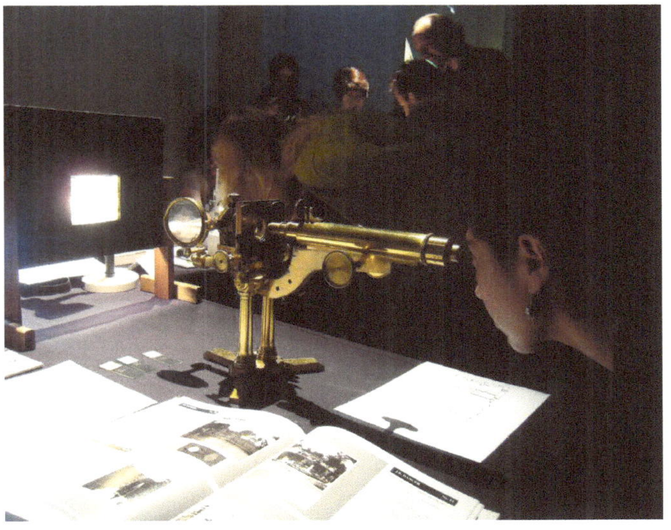

60.

55–60. Images from "Enormous Microscopic Evening," 2010
Hammer Museum, Los Angeles. Photos courtesy CRITTER Salon

RICHARD PELL
CENTER FOR POSTNATURAL HISTORY
2010

Spearheaded by artist Richard Pell, The Center for PostNatural History is a project whose objective is to advance "knowledge relating to the complex interplay between culture, nature, and biotechnology." It is an out-of-the-ordinary natural history museum that concerns itself with "PostNatural" varieties of life normally excluded from scientific taxonomy, i.e., those organisms that have been altered by humankind via selective breeding, genetic engineering, or other methods of biological tampering. Pell and his Center for PostNatural History are archivists of the biologically weird.

The Center, which keeps a catalogue of living and dead specimens, as well as publishes documents like "Strategies in Genetic Copy Prevention,"[1] celebrates the transgenic mosquitoes, fluorescent fruit flies, and inbred laboratory rats that are the by-products of our age's relentless biological experimentation. By implementing the traditional methodologies of preservation and display—insects stuck on pins, dioramas in glass cases—used by natural history museums, Pell aims to merge his practice with the larger discourse of museumology. This approach is having a great deal of success: Pell was recently awarded a research fellowship at the Smithsonian National Museum of Natural History, effectively blurring beyond recognition the border between art and science occupied by the Center for PostNatural History.

1. This publication and an accompanying installation of the same name presents a historical perspective on human inhibition of reproductive systems.

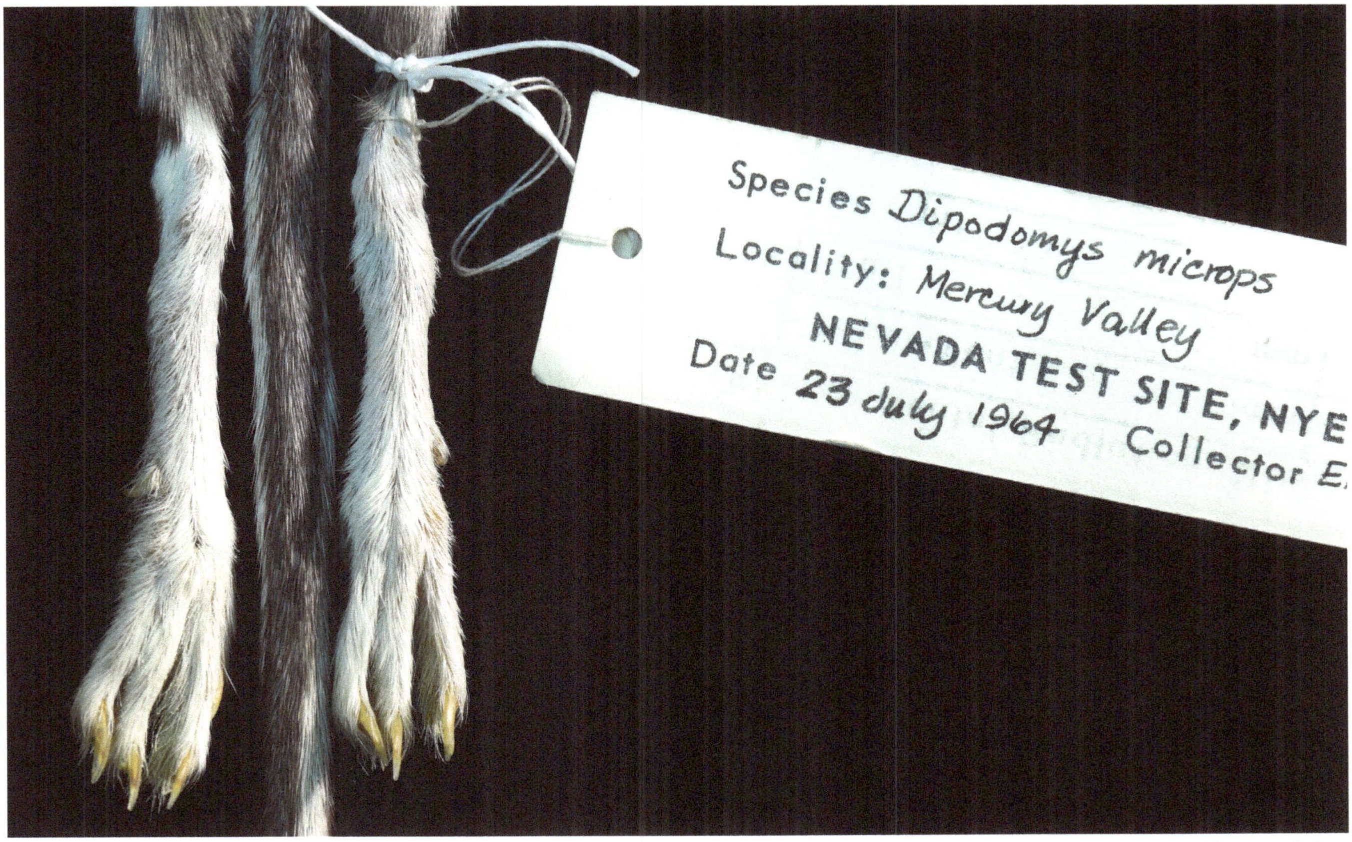

61. A small mammal collected from the Nevada Test Site following the end of atmospheric nuclear testing. Image courtesy the PostNatural History website

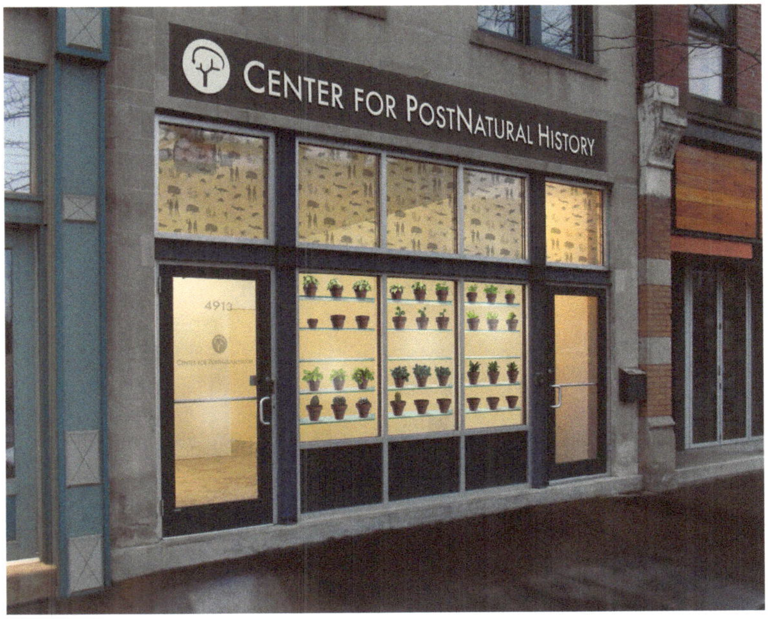

62. Rendering of the future Pittsburgh storefront to house the Center for PostNatural History, 2011

THE INSTITUTE FOR FIGURING HYPERBOLIC CROCHET CORAL REEF 2005-ONGOING

The Institute for Figuring operates in a zone between participatory science and handicraft. In 2005, as an homage to the Great Barrier Reef in their native Australia, Margaret and Christine Wertheim, founders of the Institute for Figuring, instigated a project to crochet a woolen reef. The project began in their living room in Los Angeles and has since become a worldwide movement that engages communities across the globe. *The Crochet Reef* project is a unique fusion of art, science, mathematics, handicraft and community practice that is intended to draw attention to the present environmental threats to the reef, including pollution, climate change, fishing, shipping, and human use. These images from the 2011 Smithsonian installation of *Hyperbolic Crochet Coral Reef* represent the contributions of over 1,000 participants in this collective crocheting effort.

The technique of "hyperbolic crochet" used by the Wertheims and their countless collaborators was discovered in 1997 by Cornell University mathematician Dr. Daina Taimina. As the Wertheims write on the project's website, "The basic process for making these forms is a simple pattern or algorithm, which on its own produces a mathematically pure shape, but by varying or mutating this algorithm, endless variations and permutations of shape and form can be produced."

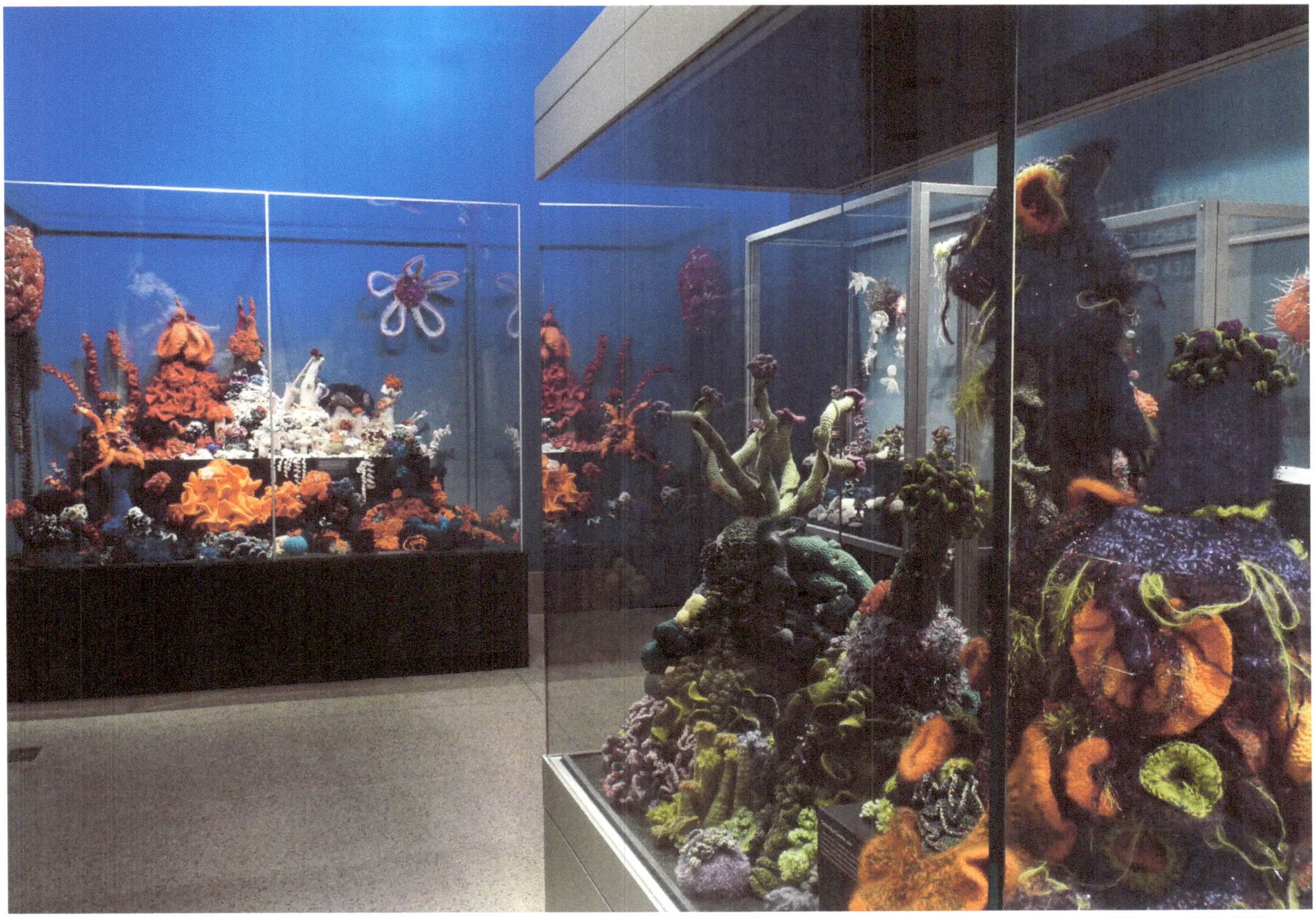

63. The Institute For Figuring's *Hyperbolic Crochet Coral Reef* project as installed at the Smithsonian's National Museum of Natural History, 2010-2011. Photo © the IFF

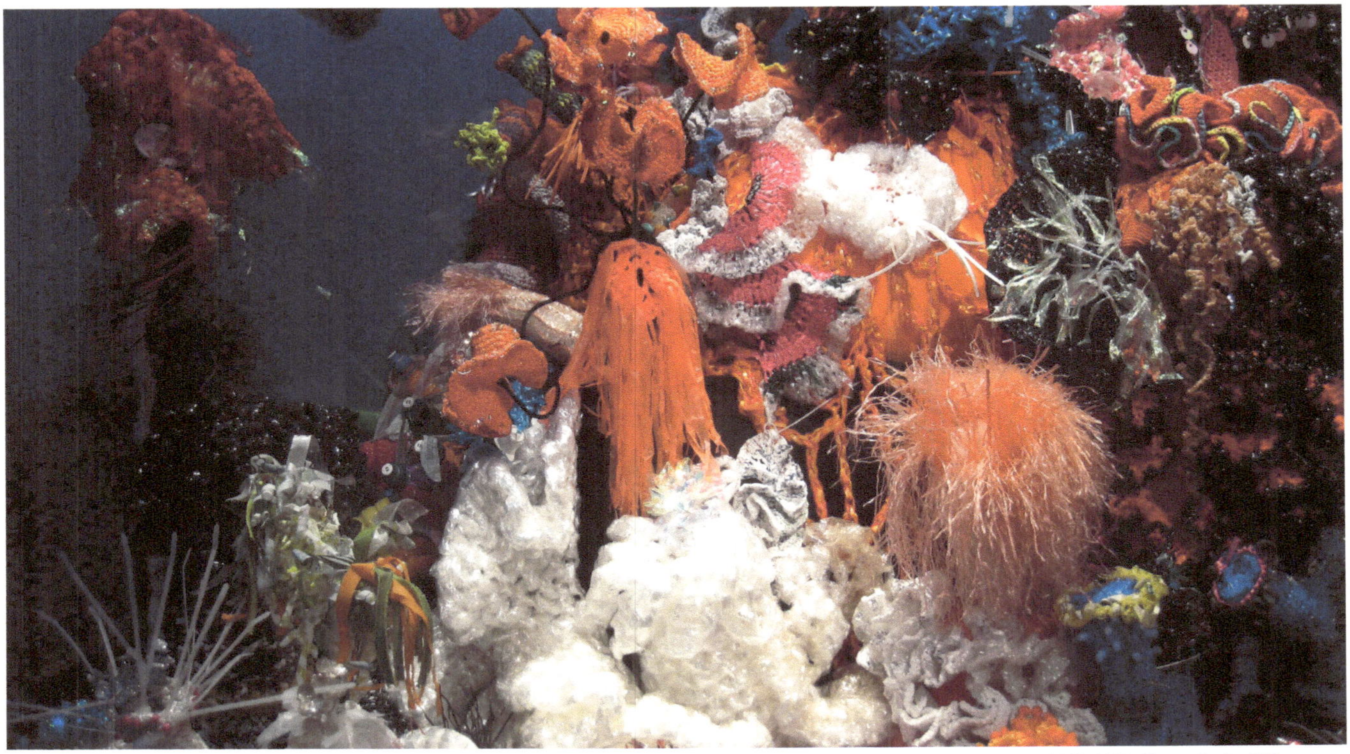

64. The Institute For Figuring's *Toxic Reef* (detail) as installed at the Smithsonian's National Museum of Natural History, 2010-2011. Photo © the IFF

Fred Adams

As a scientist, how do you see your role in a technological or scientific setting?

Much of scientific communication is done through talks, both as seminars and at scientific conferences. You might think that scientists would actually read the papers, and they do sometimes, but the spoken word carries more weight. In the old days, everyone used view graphs, but now everyone uses power-point. The art of power-point is thus vital to scientific communication, especially in astronomy where many stunning pictures (both from real data and from artists' conceptions) are available. These power-point presentations often include movies and animations. Some of these movies are generated from the results of numerical simulations (i.e., calculations that actually use the laws of physics) and sometimes they are simply animations. In the realm of "artists' conceptions," opportunities exist for future collaboration between artists and scientists, although it is difficult to find an ideal collaboration.

One could also ask how my work might actually affect technology of the future. In other words, why should my work (or any astronomy/astrophysics) be pursued? The chance of any one particular researcher making a truly important contribution (e.g., like inventing the laser, which was quite a good one) is rare. However: In order for such low probability events to arise, one has to roll the dice a larger number of times. In this context, one needs a full slate of scientific research being carried out at all times. This work should span the spectrum of topics, and should vary from extremely "applied" or "practical" (e.g., weather prediction) to the complete opposite (e.g., string theory). With all of these scientists carrying out research over this wide range, some groups or individuals will hit upon "the next big thing." Also: in order for the big breakthroughs to take place, one often needs a number of much smaller results to push them forward. These smaller contributions are not as well-recognized (you hear about Einstein a lot more than Lorentz) but are nonetheless crucial. I am happy to be able to take part in this grand enterprise. On the other hand, my own work thus far has been rather modest on the 'practical' scale: I have contributed to our understanding of how stars form and how the future universe will unfold, and my results on both of these topics have been put in (some) textbooks. Although these issues help us understand our universe, and our place within it, they do not help us build a better mousetrap. Some of my mathematical work on stochastic differential equations will (probably) be applicable to future technology, but the coupling is weak.

Fred Adams is an American astrophysicist. He is professor of physics at the University of Michigan and co-author of The Five Ages of the Universe, which discusses the history, present state, and probable long-term future of our universe, according to cosmologists' current understanding.

ARTISTS IN WHITE COATS AND LATEX GLOVES

LABORATORY, LIFE SCIENCES, HARDWARE, SOFTWARE, WETWARE, BIOART, DIYBIO, RESEARCH, VULGARIZATION, DEMOCRATIZATION

Artist researchers work in and across scientific disciplines while maintaining a creative approach. In this practice, art and science, although seemingly culturally divergent, attempt to inform one another. The objective is not to acquire expert knowledge, but rather to achieve experiential knowledge. However, many of the artists detailed in this chapter attempt to maintain legitimacy at a scientific level by adhering to the rigorous methods and standards followed by the scientific community.

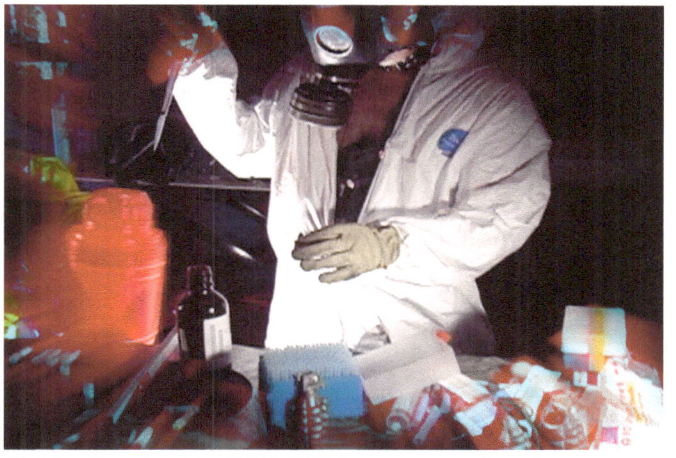

65. Adam Zaretsky with *pFARM*: 2009. Image courtesy the artist

SymbioticA[1] is often cited as the gold standard for the integration of artists inside science laboratories. Located at the School of Anatomy and Human Biology, The University of Western Australia, SymbioticA is the first research laboratory where artists and thinkers can participate in practices such as neuroscience, molecular biology, anatomy, and ethics under the umbrella of a science department. In addition, the laboratory offers researchers a chance to freely engage

1. http://www.symbiotica.uwa.edu.au/

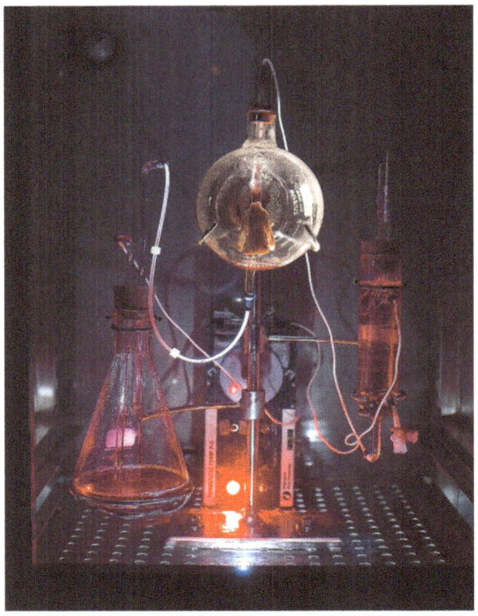

66. The Tissue Culture & Art Project, *Victimless Leather (Prototype)*, 2004. Photo: Ionat Zurr

with explorations that might not fit into the current culture and curriculum of scientific research.

Although many works developed by artists within SymbioticA's protocols might seem to be subversive, speculative, or even chimerical, they still must fully comply with the rules and requirements of scientific research. This rule-abiding approach makes their work all the more powerful and gives the artists more freedom to create and exhibit their work without the fear of being censored or excluded from the larger departmental discourse.

During a presentation at Ars Electronica in 2007,[2] Professor Stuart Bunt, one of the co-founders of SymbioticA, explained the contributions that artists can offer to science: "In science you have to work towards an end point, to 'cure,' it's not about doing research anymore, scientists are problem solvers. The critical edge that artists bring help scientists justify and constantly evaluate the scientific process."[3]

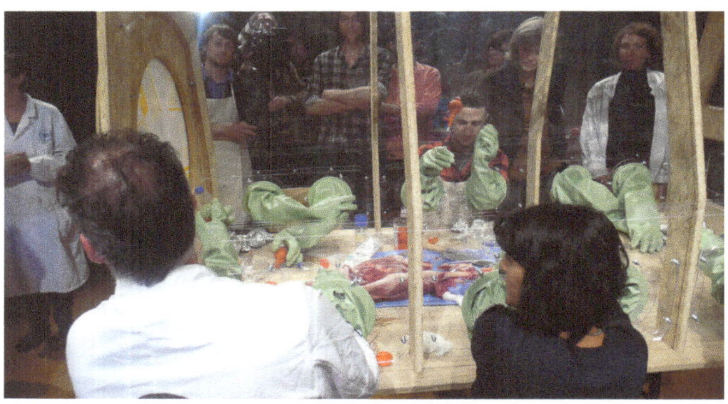

67. Vastal workshop with Adam Zaretsky and Oron Catts at the Waag Society in Amsterdam, September 2009. Photo: Régine Debatty

One of the key purposes of artists' explorations and experiments is to bring scientific discussions out of the laboratory and to initiate a public discourse about not only the state of scientific research but also its ethical, social, political, and cultural implications. As the late Professor Stephen Wilson[4] noted, "Technology/science/art research is still marginalized as a fringe activity. In a technoscientific culture, artistic probing of the world of research is a critical, desperate need. We need people looking at these fields of inquiry from many frames of reference, not just those sanctioned by academia or commerce."[5]

Some artist researchers also make it their mission to disseminate knowledge and make technology more accessible

2. http://90.146.8.18/en/festival2007/program/schedule.asp
3. http://we-make-money-not-art.com/archives/ars_electronica_2007/
4. Stephen Wilson was an artist, author, and professor at San Francisco State University who passed away in January 2011. He was considered to be a leading thinker and practitioner in the field of new media.
5. http://we-make-money-not-art.com/archives/2007/04/interview-with-12.php

to the general public. They invite the public to participate in hands-on workshops that explore science in a non-intimidating way. This heuristic approach to the tools and methods of science and technology not only provides participants with a more informed, practical perspective, but also acts as the starting point of a series of critical discussions about the broader cultural and ethical implications of the technology at stake.

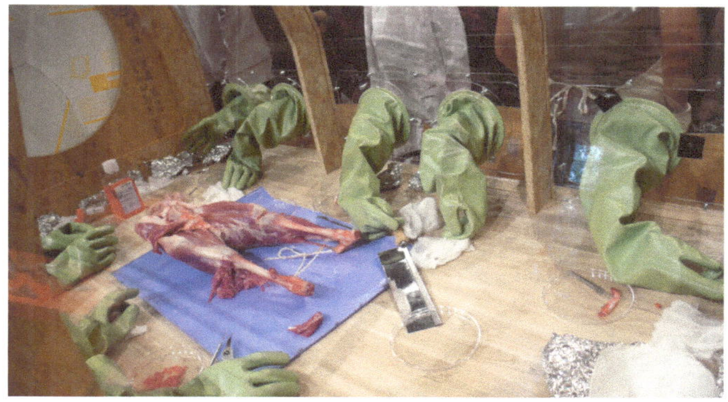
68. Vastal workshop with Adam Zaretsky and Oron Catts at the Waag Society in Amsterdam, September 2009. Photo: Régine Debatty

An example of this is artist Adam Zaretsky's VivoArts Workshops, courses that combine studio arts and biological sciences. Zaretsky explains: "The difference between a technical scientific learning session and a Vivo-artistic laboratory approach is mostly qualitative. While engaging in the technics, we also deal with the relational issues surrounding this type of process: pain, death, responsibility, curiosity, the meddlesome sadism of a personal genetic footprint/signature/graffiti, risk assessment between foreign species and the ecosphere, as well as critiquing admonitions against the urge to fondle the folds of mutant love."[6]

It is in launching these types of creative public laboratories that artists extend scientific discourse into the social sphere. This variety of public engagement happens regularly at the recently opened Science Gallery at Trinity College, Dublin. This "world's first" institution is a laboratory-like art space that regularly supports hands-on experiments, conversations, and playful social events as part of its art/science exhibitions. Typical Science Gallery events include "Seed Dating," an attempt at matchmaking between artists and scientists for future projects, and exhibitions like "Infectious" (2009), which cast a large net to understand all that is "contagious"—from laughter to viruses. Included in

6. http://we-make-money-not-art.com/archives/bioart/index.php?page=3

this exhibition was a project titled *Kiss Culture* (2009), by Maria Phelan of Ireland, in which visitors, on entry to the gallery, kissed an individual petri dish, which was then cultured so they could come back to "discover the natural flora you carry on your lips and nose."[7] It was a cultural activity in more ways than one.

7. http://www.sciencegallery.com/infectious-stay-away/infectious-exhibits/kiss-culture

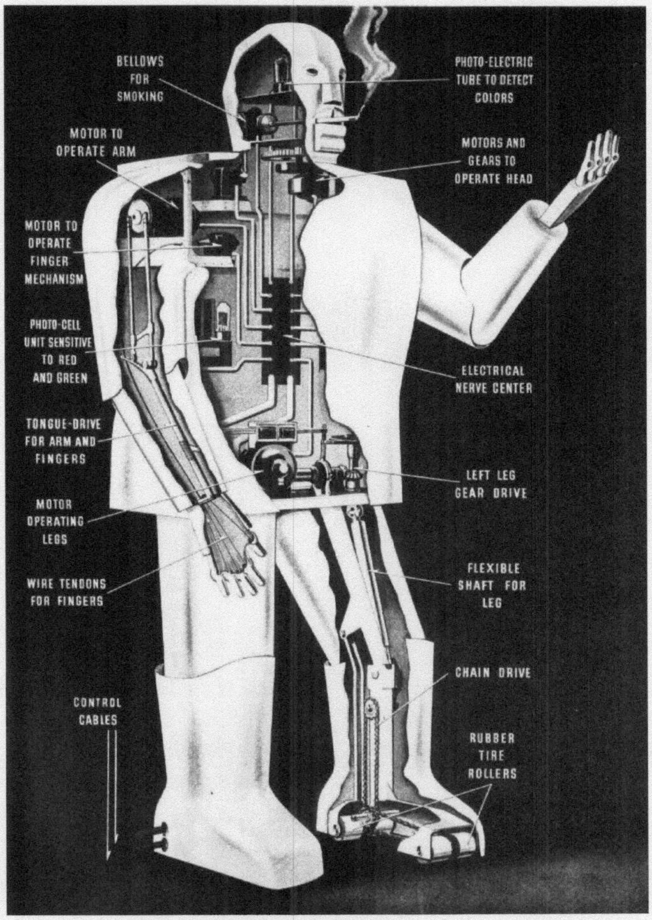

69. Elektro the Motoman, 1939 World's Fair

BRANDON BALLENGÉE
MALAMP
1996–2008

Since 1996, the biologically inclined artist Brandon Ballengée has been studying declining populations and deformities in amphibians. These investigations, conducted with scientific rigor, have involved collaborations with numerous researchers throughout the United States and Europe. According to Ballengée, amphibians are environmental canaries in the coal mine, and the state of this group of animals is dire—not only are they declining in population across the globe, but they are displaying increasing levels of morphological anomalies such as extra, deformed, or missing limbs.

Ballengée sees his research as an effort to blur the "already ambiguous boundaries" between environmental art and ecological research; his work, which he often conducts for organizations like the United States Geological Survey's North American Reporting Center for Amphibian Malformation (NARCAM), stands up in terms of both scientific methodology and aesthetic value. His work is collected by natural history institutions like the Peabody Museum at Yale University and the American Museum of Natural History in New York, as well as contemporary art museums and galleries. In his work Ballengée could be considered a descendant of the 19th century Hudson River School of landscape painting and of 20th century Land Artists like Betty Beaumont and Agnes Denes. A practitioner of fieldwork and color, Ballengée is as comfortable in a laboratory as he is in a lecture hall—a twenty-first-century polymath.

All images contain information of some kind. Traditionally, "art" images provide visual, conceptual, or aesthetic information that stimulates us and moves us in specific ways, while "science" images have a more pragmatic value, designed to convey measurements and relationships—data. Ballengée's work contains both varieties and, by being ecological in nature, is also political. Of his images, he writes that they are a "visual dialogue that become a conceptual form of environmental outreach."

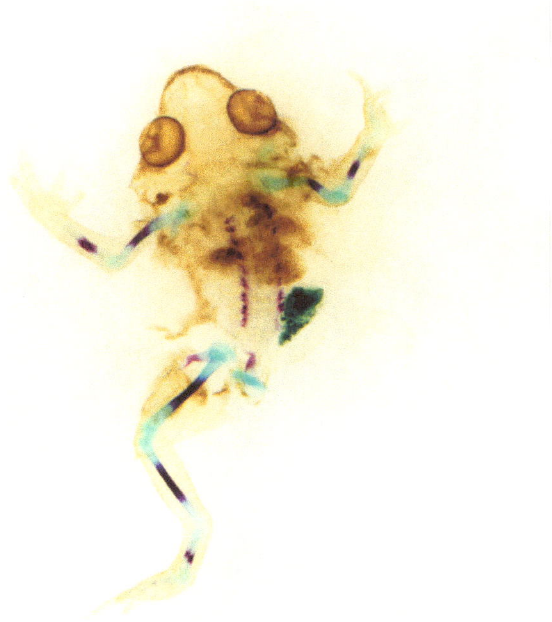

70. *DBB 2, Khaos*, Scanner Photograph of Cleared and Stained Deformed English Toad in Scientific Collaboration with Richard Sunter (Yorkshire Naturalist's Union). Unique Digital Chromogenic print on watercolor paper, 121 cm. by 93 cm., 2008/10, Commissioned by Arts Catalyst, London, England

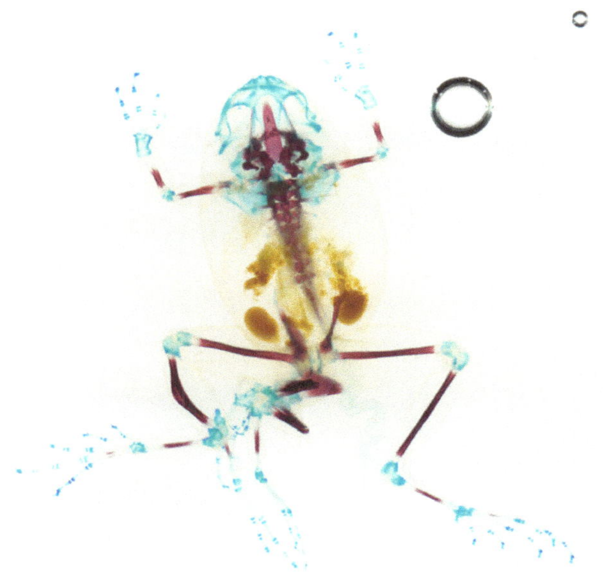

71. *DFA 8, Deméter*, Scanner Photograph of Cleared and Stained Multi-limbed Pacific Tree frog from Aptos, California in Scientific Collaboration with Dr. Stanley K. Sessions. MALAMP titles in collaboration with the Poet KuyDelair, H 46 inches x W 34 inches, 117 cm by 86 cm, IRIS print on watercolor paper, 2003/07, Courtesy the Artist and Nowhere Gallery, Milan

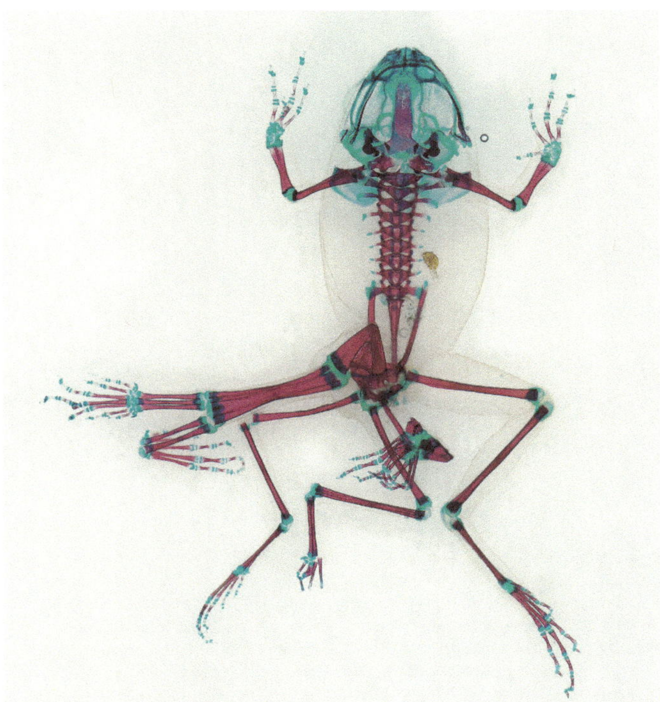

72. *DFA 18, Triton*, Scanner Photograph of Cleared and Stained Multi-limbed Pacific Tree frog from Aptos, California in Scientific Collaboration with Dr. Stanley K. Sessions. MALAMP titles in collaboration with the poet KuyDelair. H 46.5 inches x W 34.5 inches, 118 cm by 88 cm, Unique Digital Chromogenic print on watercolor paper, 2001/07, Courtesy the Artist and Verbeke Gallery, Antwerp

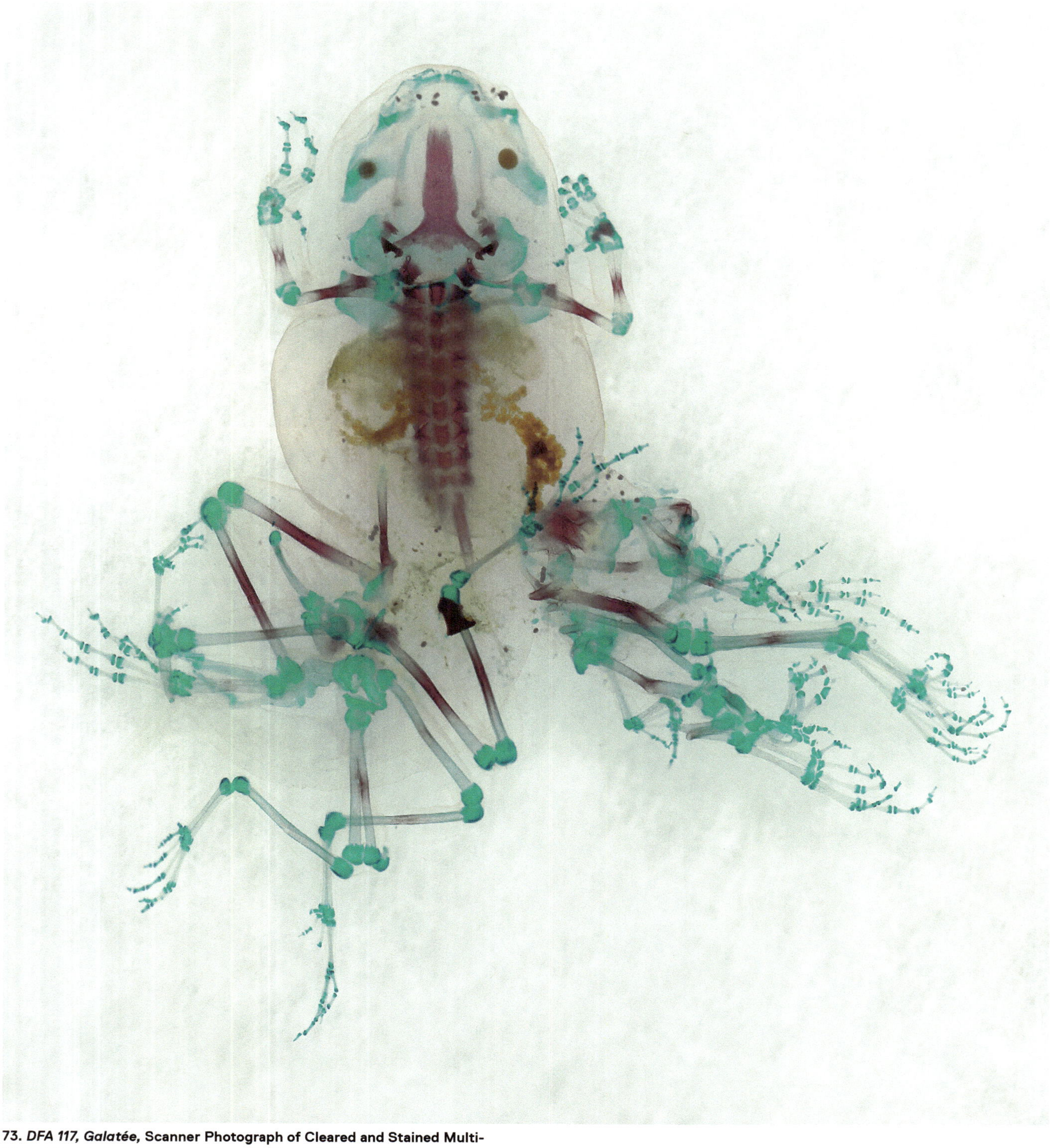

73. *DFA 117, Galatée,* Scanner Photograph of Cleared and Stained Multi-limbed Pacific Tree frog from Aptos, California in Scientific Collaboration with Dr. Stanley K. Sessions. MALAMP titles in collaboration with the poet KuyDelair. H 46.5 inches x W 34.5 inches, 118 cm by 88 cm, Unique Digital Chromogenic print on watercolor paper, 2001/07, Courtesy the Artist and Verbeke Gallery, Antwerp

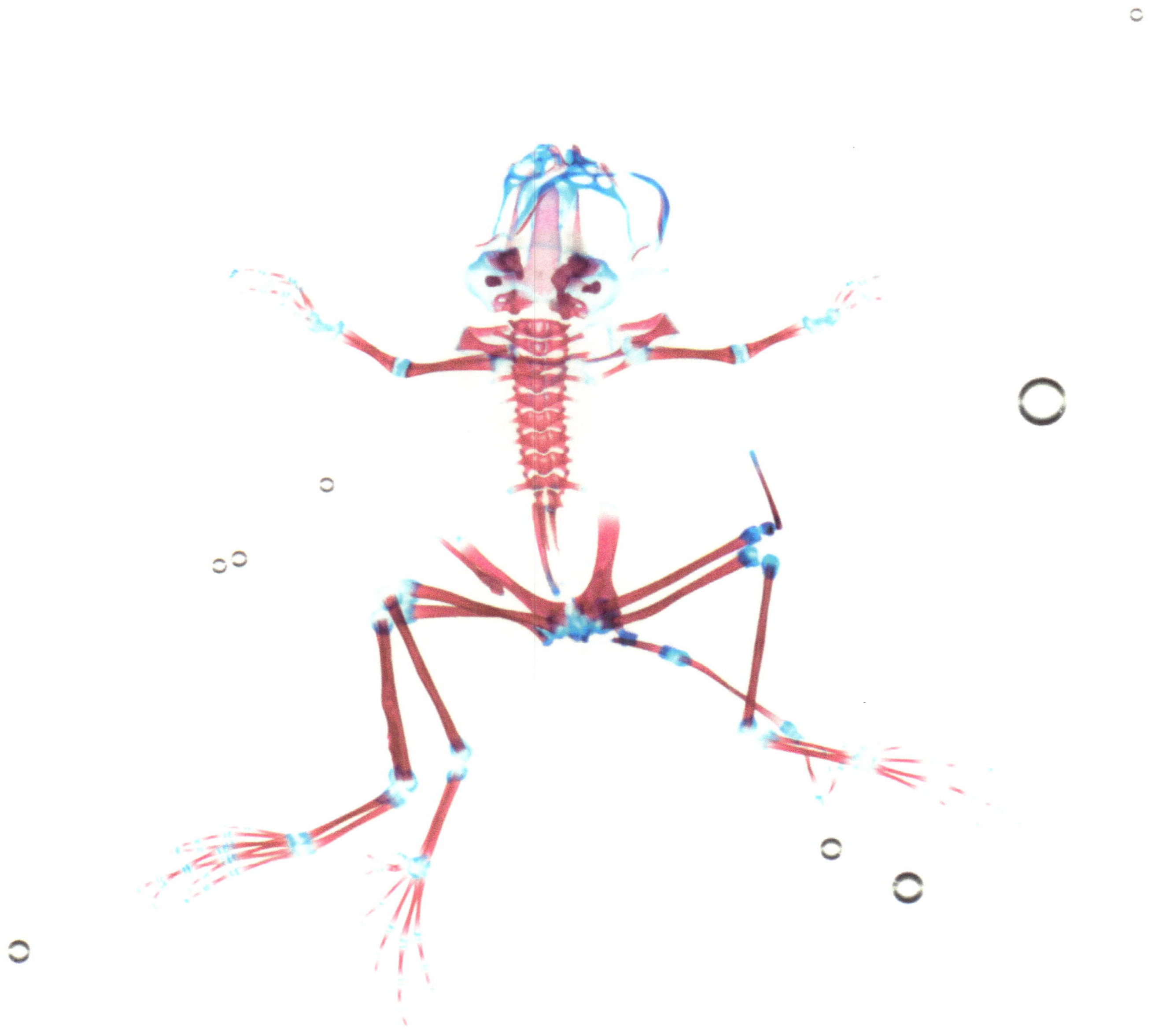

74. *DFA 9, Sphinx,* Scanner Photograph of Cleared and Stained Multi-limbed Pacific Tree frog from Aptos, California in Scientific Collaboration with Dr. Stanley K. Sessions. MALAMP titles in collaboration with the Poet KuyDelair, H 46 inches x W 34 inches, 117 cm by 86 cm, IRIS print on water-color paper, 2003/07, Courtesy the Artist and Nowhere Gallery, Milano

GILBERTO ESPARZA
NOMADIC PLANTS
2010

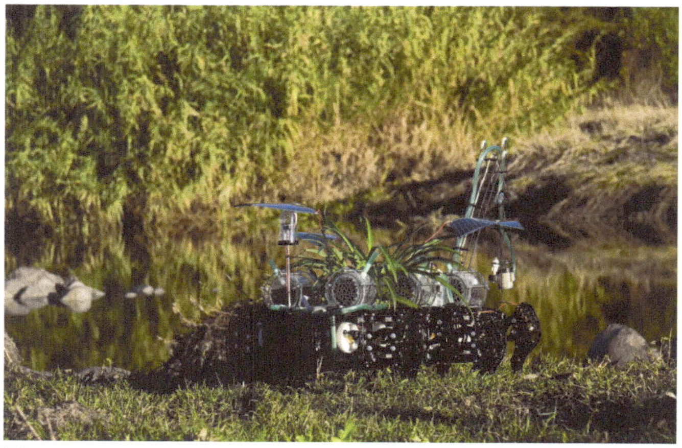

75.

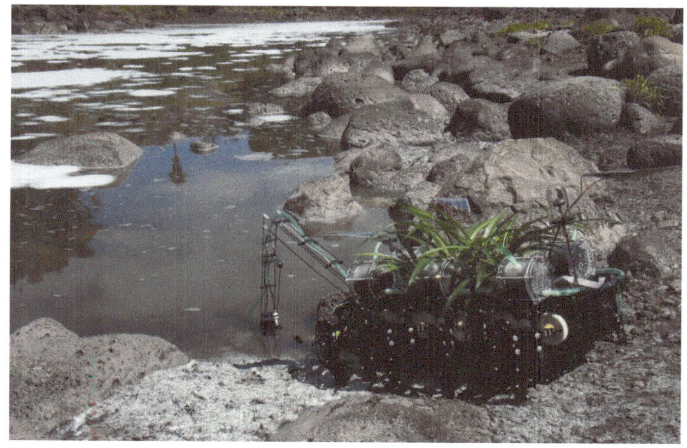

76.

75, 76. Images from Gilberto Esparza, *Nomadic Plants*, 2010. Images courtesy the artist

The *Nomadic Plants* created by Gilberto Esparza are robots powered by water pollution—the dirtier, the better. Whenever the plants and microorganisms living in symbiosis inside the robot's body require nourishment, the robot moves toward a contaminated body of water and "drinks" from it. As the water passes through a microbial fuel cell,[1] the toxic elements in the water are decomposed and turned into energy that can feed the robot's "brain circuits." Any surplus water is then used to sustain plants living on the robot's back.

"Nowadays robots are a waste of energy," Esparza told *Icon Magazine*. "They dance and they move all the time."[2] *Nomadic Plants*, on the other hand, are both efficient and autonomous.

Esparza uses his art to explore new ways of engaging with current ecological issues, illustrating the importance of living in symbiosis with our planet and all species that inhabit it. His *Nomadic Plants* project is part of a series of experiments that aims to stimulate a critical discussion about the ambiguous forces wielded by technology.

"The Nomadic Plant is a portrait of our own species," explains the artist. "It also deals with the alienated transformation of this new hybrid species that fights for its survival in a deteriorated environment."[3]

1. Microbial fuel cells use bacteria to achieve the direct conversion of organic matter to electricity.
2. http://www.iconeye.com/index.php?option=com_content&view=article&id=4324:nomadic-plants-by-gilberto-esparza
3. http://www.we-make-money-not-art.com/archives/2010/04/-1-cuando-lei-acerca.php

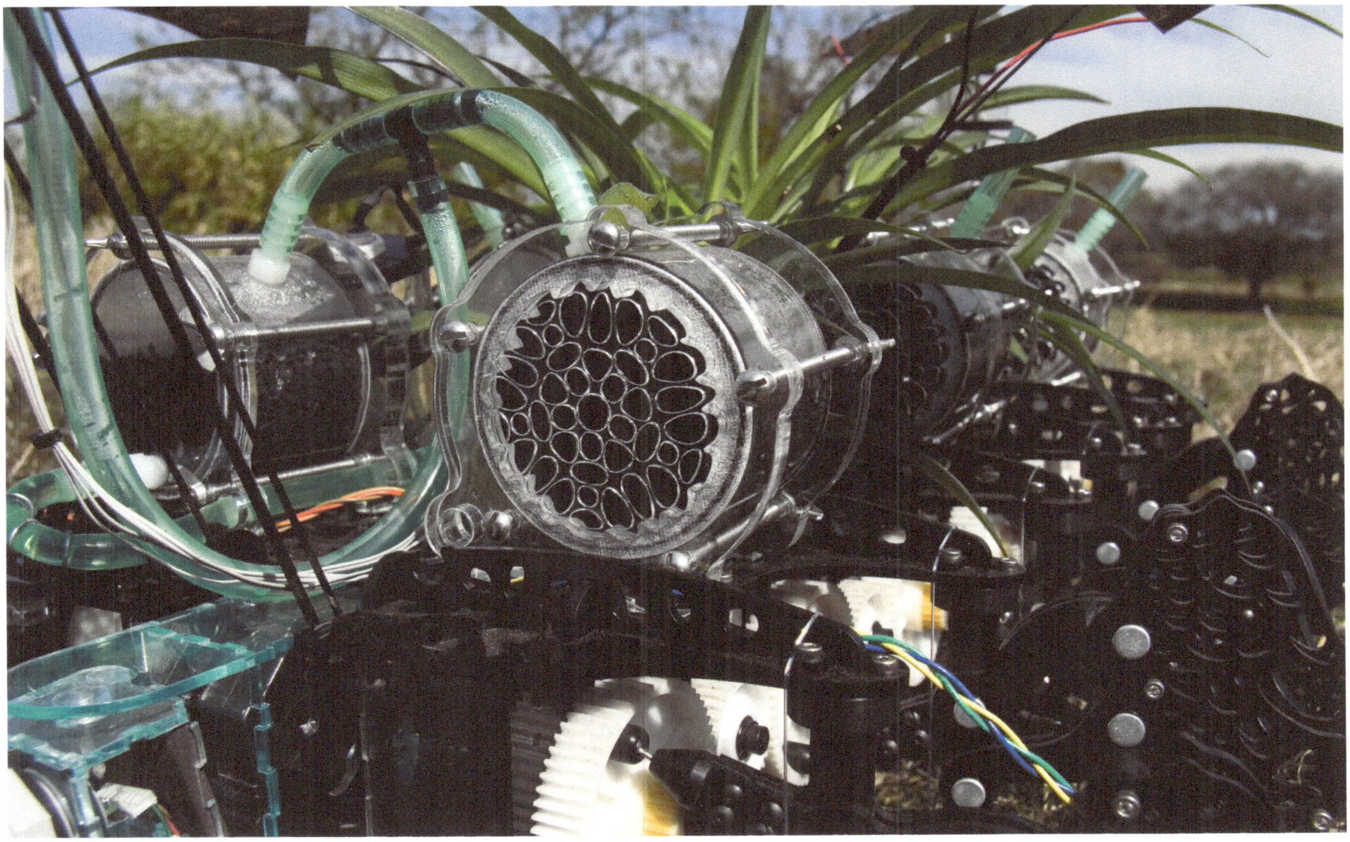

77.

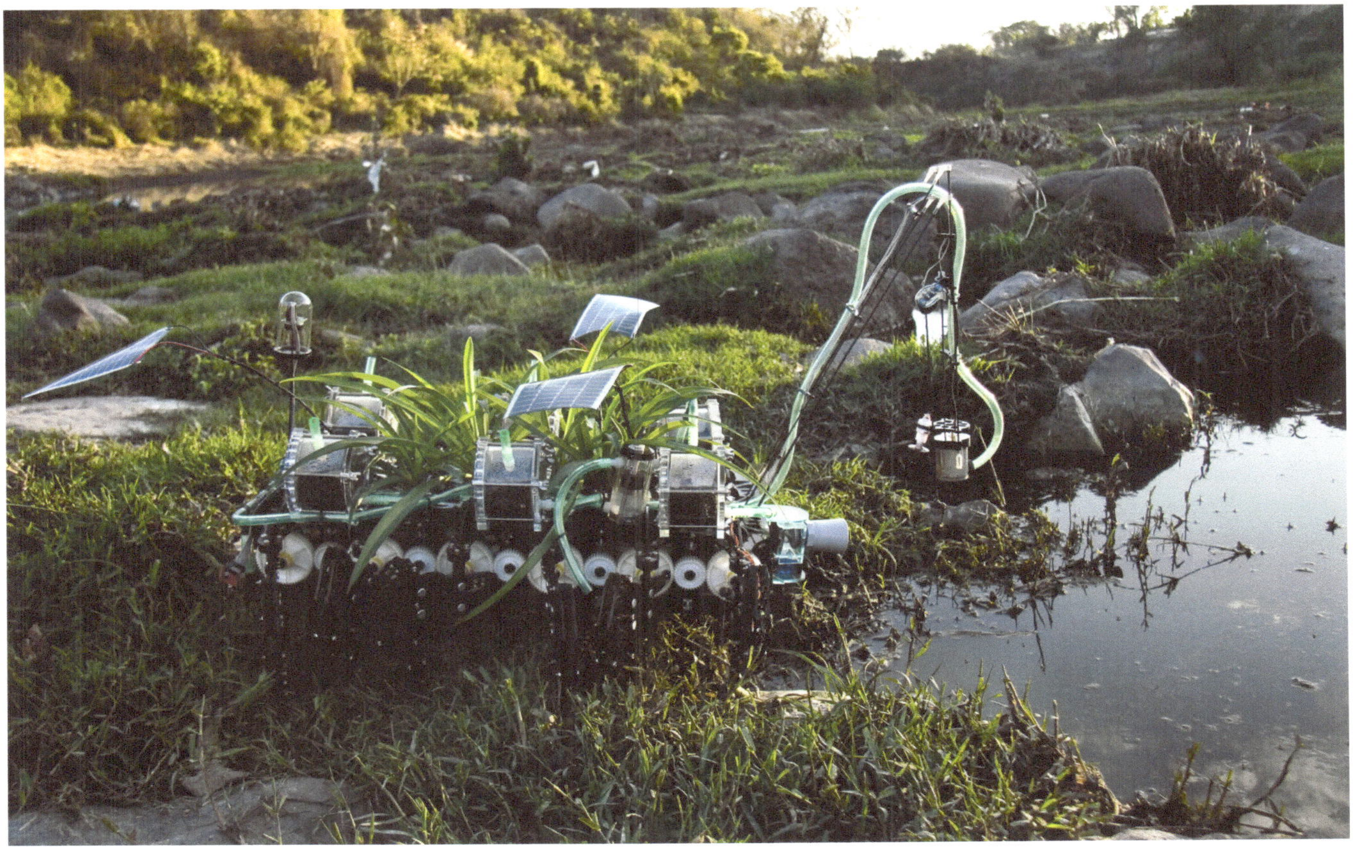

78.
77, 78. Images from Gilberto Esparza, *Nomadic Plants*, 2010. Images courtesy the artist

PHILIP ROSS
MYCOTECTURAL ALPHA
2009-ONGOING

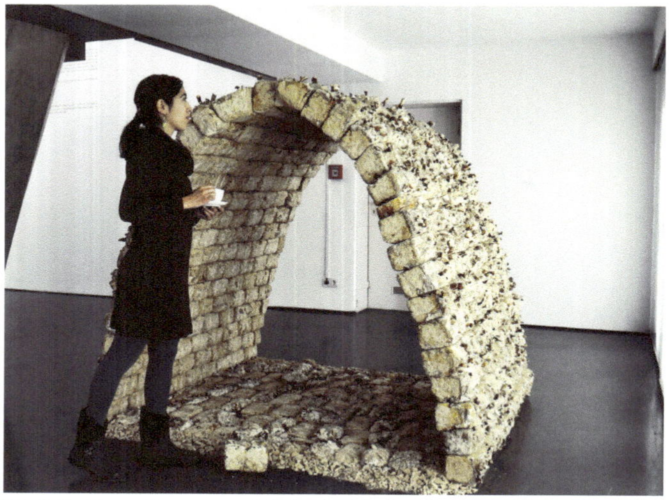

79. Philip Ross, *Mycotectural Alpha*, installation view, 2009

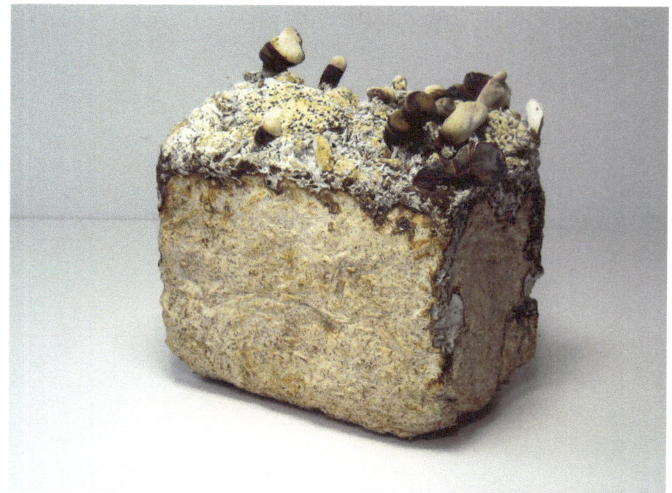

80. A single *Ganoderma lucidum* brick from Philip Ross, *Mycotectural Alpha*, 2009

Philip Ross's *Mycotectural Alpha* was a catenary arch grown from the fungus *Ganoderma lucidum* (also known as Reishi or Ling-chi). The fungal bricks, grown in wooden cases over the course of several months at a mushroom farm in Northern California, look like, in Ross's words, "decrepit cake." Although the piece appeared whimsical, the idea of building structures out of mushrooms is not as far-fetched as one might think. Mycelium, the network of thin, rootlike fibers that forms beneath a mushroom, is non-toxic, is resistant to fire, mold, and water, and traps heat better than fiberglass insulation does. It's also possibly the only material on Earth to be both stronger than concrete *and* edible.

Mycotectural Alpha was the first structure made entirely out of mushrooms. According to Ross, it was so sturdy that he damaged countless metal files and saw blades in the process of constructing the arch. He writes, "My goal is to create a space that can shelter 12-20 people at a time. Over the next few years I will continue growing experiments to determine the fungi's material qualities as well as a means of propagating more complex forms."[1]

The arch, which was a test structure preceding Ross's further experiments in "mycotecture," was boiled down into tea over the course of its installation at the Kunsthalle Düsseldorf in 2009. "A literal tea house," Ross writes.

1. This and all other quotes are from http://www.philross.org/.

PHILIP ROSS
PURE CULTURE
2007-ONGOING

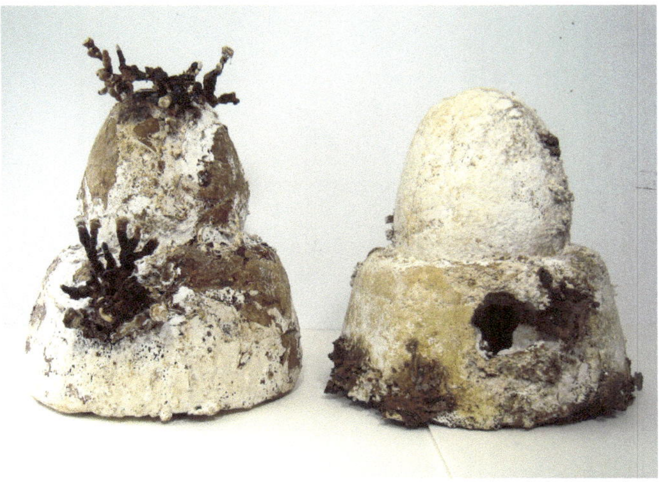

81.

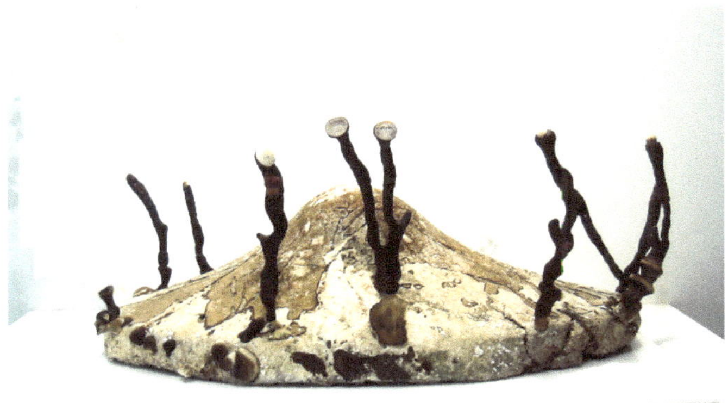

82.

81, 82. Philip Ross, *Pure Culture*, 2007-ongoing

Ross sees fungal material as having potential as architectural building blocks, and he's not alone in this belief. In a case of art and science synchronicity, a young company called Ecovative, with a myco-factory in upstate New York, produces and sells "Greensulate," an insulation and packaging material made from mushroom mycelium. Ecovative, which calls this material a "green alternative to Styrofoam," has been awarded grants from the EPA, National Science Foundation, and Department of Agriculture.

However, not all of Ross's mycological work is concerned with function. Since 2007, he has been working on a series called *Pure Culture*, comprising mushroom sculptures of a purely formal nature; for these, he molds *Ganoderma lucidum* into esoteric and fantastical forms over the course of months and even years of growth in his studio. Ross compares his mushroom sculptures to bonsai—a practice of "transforming and refining."

Philip Ross

As an artist, how do you see your role in a technological or scientific setting?

There are various roles I believe I play in technological and scientific environments, some of which include:

— Being a translator of scientific method through educational programs and cultural activities.
— Acting as an instigator of dialogue between the public, academies and other centers of informational power and authority.
— Wielding the axe of the barbarian in an attempt to blow open the civilized gates, doors and walls that deny access to information.
— Learning through the actions of a symbiont, and exchanging resources and knowledge in relationships that are mutually beneficial to those in participation.
— Producing critical artworks and writings that reflect my interest in the history, poetics and environments of practical knowledge.

My evolution into these roles came about through growing up with an intensive interest in rationalized aesthetics and other toys of modernity, but even more importantly through the desperate need to understand the literacy of power that governs critical medical and health decision making processes. What follows is biographical, but is the best way I know to explain the source of inspiration for being involved so intensively with science and technology.

I moved to San Francisco in the mid 80's to attend the San Francisco Art Institute, in part because the Bay Area had a thriving techno culture as well as an equally vibrant techno counter-culture. This was the place that was home to Ant Farm and Terry Riley. AutoDesk was developing those weird data gloves and virtual ski goggles, and SRL was putting together political killer-robot sex wars. From the distance of the East Coast where I grew up this all seemed spectacularly miraculous. I wanted my turn at flying around in the chrome plated UFOs that I imagined all the citizens of Cupertino and Redwood City were given the keys to at birth.

Of course, things were different than I had imagined. When I arrived the A.I.D.S. crises was in full bloom, sniffing out and destroying scores of young classmates, friends, and friends of friends. Information about the disease was hard to come by and even harder to understand. Viral mechanics are difficult to comprehend without a grasp of molecular and cellular dynamics, and understanding retroviruses requires knowledge of the immune system, RNA transcription and many other complicated biological systems. I became a hospice caregiver for friends and family, and was radicalized through participation with Act Up and other underground health advocacy networks to challenge institutional authority. The central axioms being:

Information=Power
Silence=Death

In trying to find information about existing medicines and therapies I learned about the American health care system, the statistical methods of clinical trials, and the difficulties of finding and interpreting significance in scientific language and writing. In combing through medical texts and libraries I became a true academic, and as a result was able to formulate challenges to the decisions of politicians, doctors and other 'important' people who guarded access to knowledge and other rationales of choice.

In this USA most scientific literacy ends at 10th grade if a student is not good at science or math. This was certainly the case for me. This ruins the possibility of making well-informed health decisions, often conducted at desperate times, without advocacy or a realization that there might be other opinions and possibilities that are counter to what is being offered as a best decision. I have a fire inside, and feel that our institutional and educational systems fail us as a citizenry when the wealth of our collective knowledge is cryptically removed because of lack of access, language and practice. One of my roles as an artist is to use culture and creativity to bring people into a more informed and critical relationship to technological and scientific environments.

BCL (GEORG TREMMEL AND SHIHO FUKUHARA) COMMON FLOWERS / WHITE OUT 2010

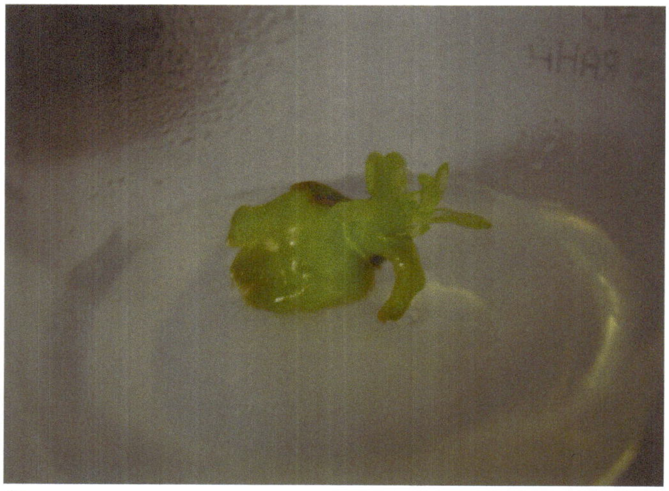

83.

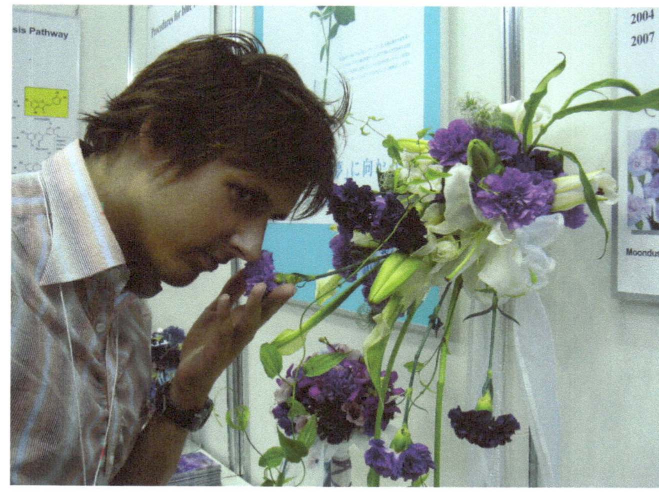

84.

83, 84. Images from BCL (Georg Tremmel and Shiho Fukuhara), ***Blue Carnation*****, 2009**

For *Common Flowers*, the artists Georg Tremmel and Shiho Fukuhara reverse-engineered a genetically modified variety of carnation called "Moondust." Moondust was designed by Suntory Flowers, which genetically manipulated an originally white carnation to produce blue flowers. It was the first commercially available, genetically engineered consumer product that was intended purely for aesthetic consumption.

BCL—the name of Tremmel and Fukuhara's practice—purchased commercial Moondust flowers and proceeded to "biohack" them in their kitchen, cultivating the cut flowers into plants. Taking the project a step further, they exploited the so-called non-harmful designation of these genetically modified plants and seeded them into the wild, creating a "Flower Commons," a free population of prospering flowers. BCL refers to the artists' liberation of a genetically engineered plant in technological terms—like a hacked cellular phone, the Moondust was "jailbroken." This work raises questions about intellectual property and copyright in the realm of nature, as the propagation of these flowers in the environment constitutes a violation of Suntory's copyright.

Whether you call it reverse bio-piracy, do-it-yourself biotech, or science fair project, BCL's *Common Flowers* is an act that establishes the empowering possibilities of biotechnology in the artistic and public lexicon. "As a cultural practice," Fukuhara explained to *The Japan Times*, "we believe it is important to gain an understanding of biotechnology. As the twentieth century was the century of the computer, the twenty-first century will be about biotechnology."

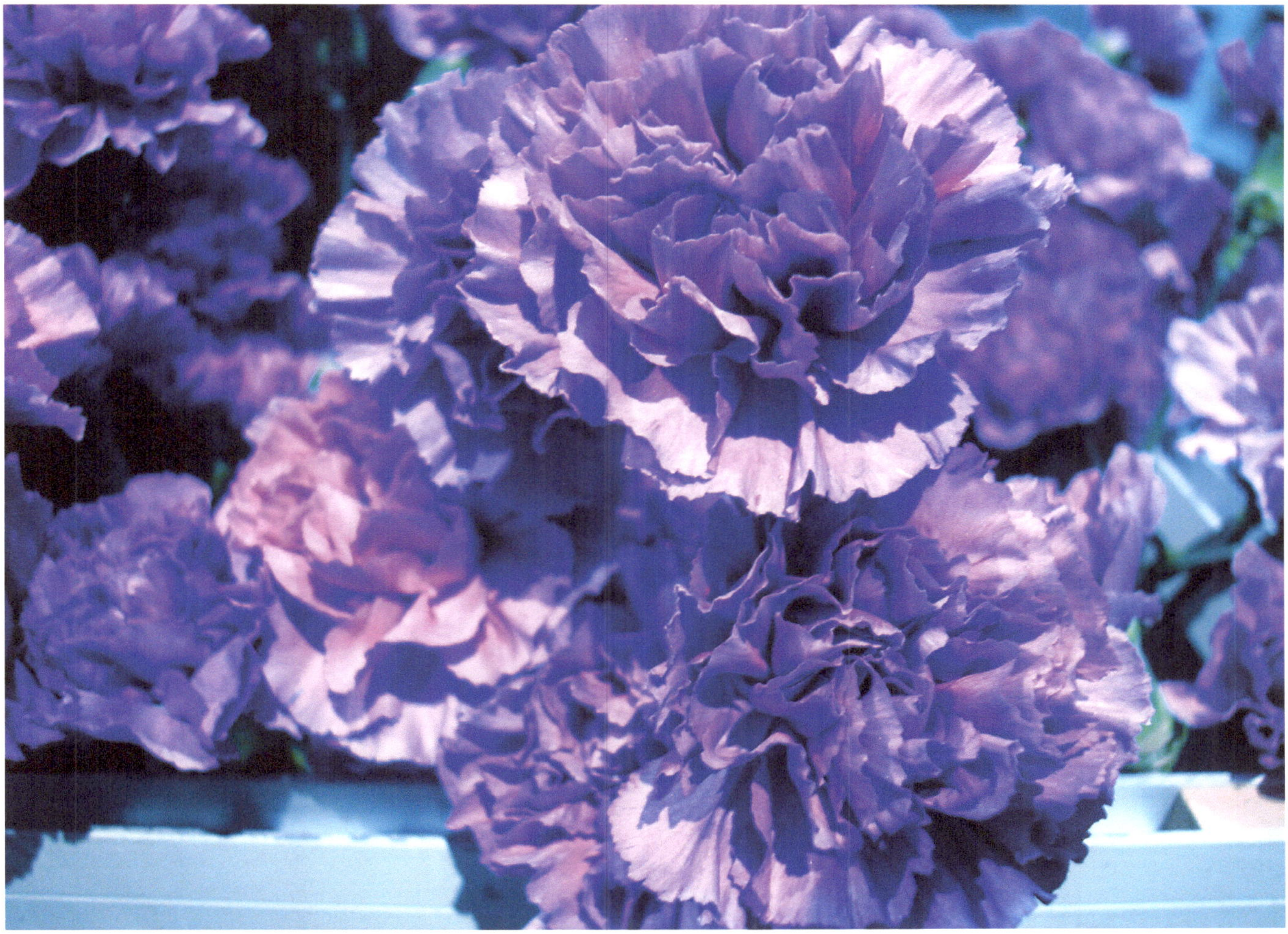

85. BCL (Georg Tremmel and Shiho Fukuhara), *Blue Carnation*, 2009

KATHY HIGH
BLOOD WARS
2011

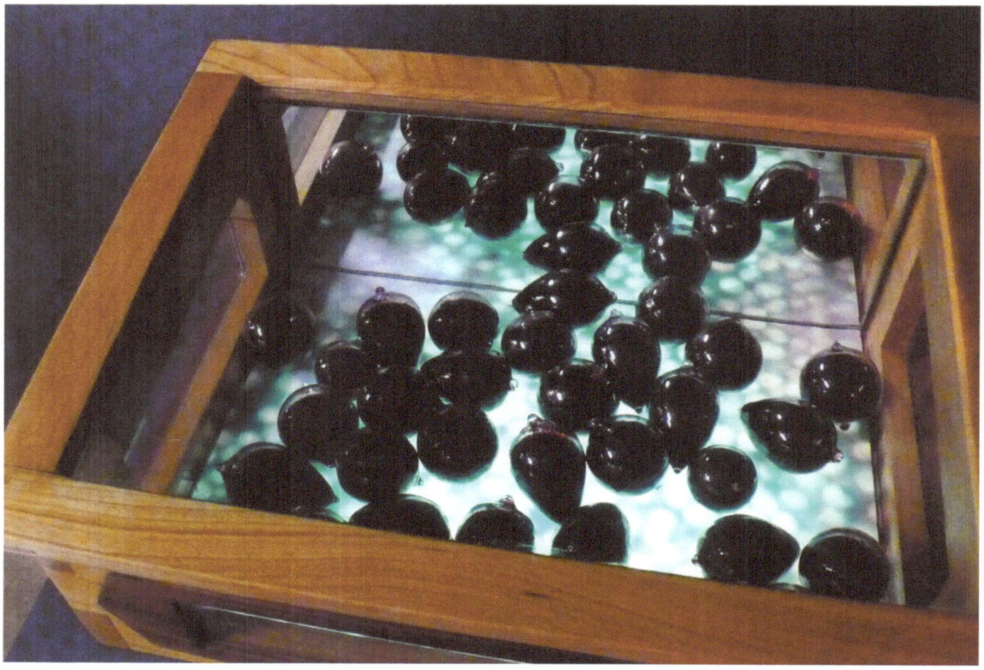

86. Kathy High, *Blood Trophies Cabinet*, 2011

87. *Blood Wars* logo

Blood Wars is a project being conducted by Kathy High, a self-styled new media artist and professor at Rensselaer Polytechnic Institute. *Blood Wars* is essentially a tournament between different donors' white blood cells held in a petri dish.

First, human blood samples are drawn by a professional phlebotomist. The white blood cells are isolated, stained for identification, and placed inside a petri dish for the duel, which pits blood cells from different participants against each other to compete for nutrients in their environment—an uncommon encounter. The cellular "winner" of each match, which is photographed under a microscope, goes on to fight the next participant.

Using a questionnaire describing the identity of each participant, High gathers this information to generate a fictional character for the individual blood samples—each "blood type." The project explores the possibilities presented by this strange medium, delving into the character of the human immune system and the ways we think about the the body's "defenses" in bio-political terms. *Blood Wars* is also an artistic effort to question some deeply entrenched ideas of racial superiority, the idea of having blood-based kinship, and how our blood has traditionally defined our sociopolitical and racial identity (i.e., "blue blood").

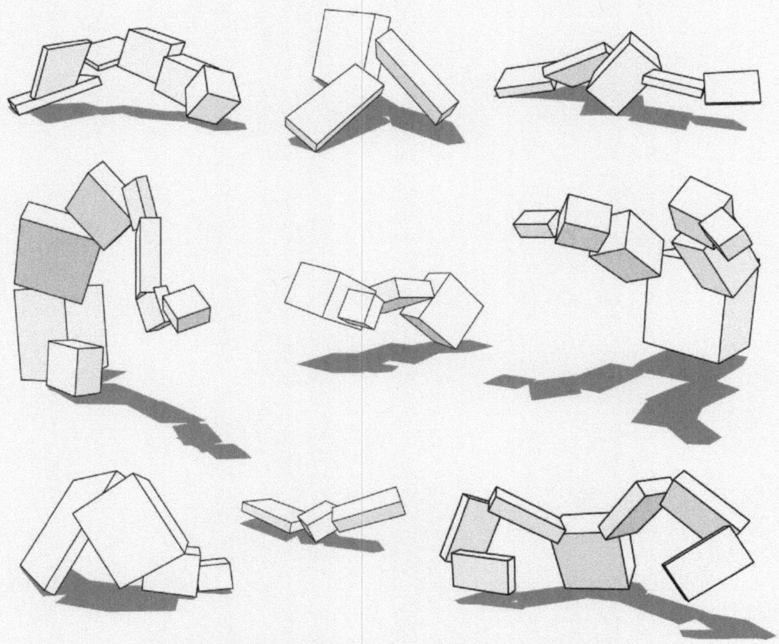

Figure 7: Creatures evolved for walking.

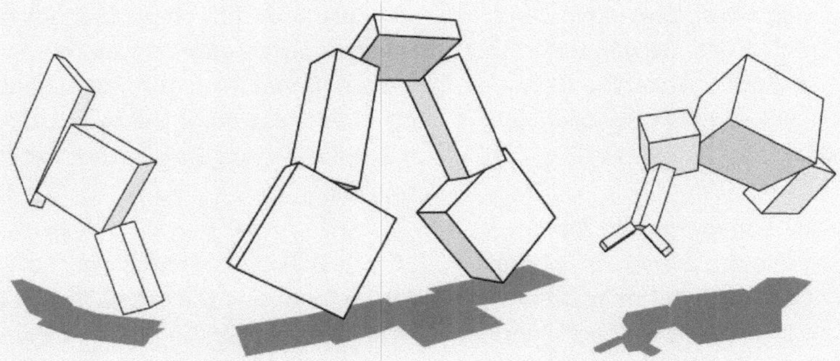

Figure 8: Creatures evolved for jumping.

88. Karl Sims, *Evolving Virtual Creatures*, 1994

FERNANDO ORELLANA
SLEEP WAKING
2008

Sleep Waking, a collaboration between artist Fernando Orellana and computer scientist Brendan Burns, investigates the human-robot relationship. Using brainwave activity and eye movements recorded during REM sleep to generate robot behaviors and head positions, the artist enables the *Sleep Waking* robot to "play back" human movement theoretically occurring during the dream state.

To develop the piece, Orellana spent a night at the Albany Regional Sleep Disorder Center in Albany, New York, where he was wired up with a variety of sensors that monitored everything from his brain using an EEG (electroencephalogram) to his heart using an EKG (electrocardiogram), as well as eye positioning data. The data collected on Orellana's eye movements was applied to direct the position of the robot head, while the EEG data was run through a machine-learning algorithm to identify patterns, which were then associated to pre-programmed robot behaviors. The patterns were used as filters to process the entire data set, and the robot acted out each associated behavior accordingly. Periods of high REM activity were associated with dynamic behaviors in the robot—pretending to fly, for example, or looking scared. Low REM activity, on the other hand, triggered more subtle behaviors—like gesturing and looking around. This range of behaviors demonstrated by the robot correlate to actions a human might perform in a dream.

Robots are increasingly being used to augment human experience. From prosthetic devices and robotic telepresences to implanted RFID chips, technology is advancing robots from being an externalized tool to being a literal extension of who we are. Some form of robot-human hybridity seems to be an inevitable feature of the near future, and Orellano's work gives examples of, and draws attention to, the paths we take toward this end.

"*Sleep Waking* is a metaphor for a reality that could be in our future," explains the artist. "In the piece we use a fair amount of artistic license. Though the eye positioning data is a literal interpretation, what we do with the EEG data is a bit more subjective. However, perhaps one day we will have the technology to allow a robot to act out what we do in our dreams. What could we learn from seeing our dreams played back for us? Will we save our dreams [the way] we save our photographs?"[1]

1. http://we-make-money-not-art.com/archives/2008/02/how-does-it-work-exactly.php

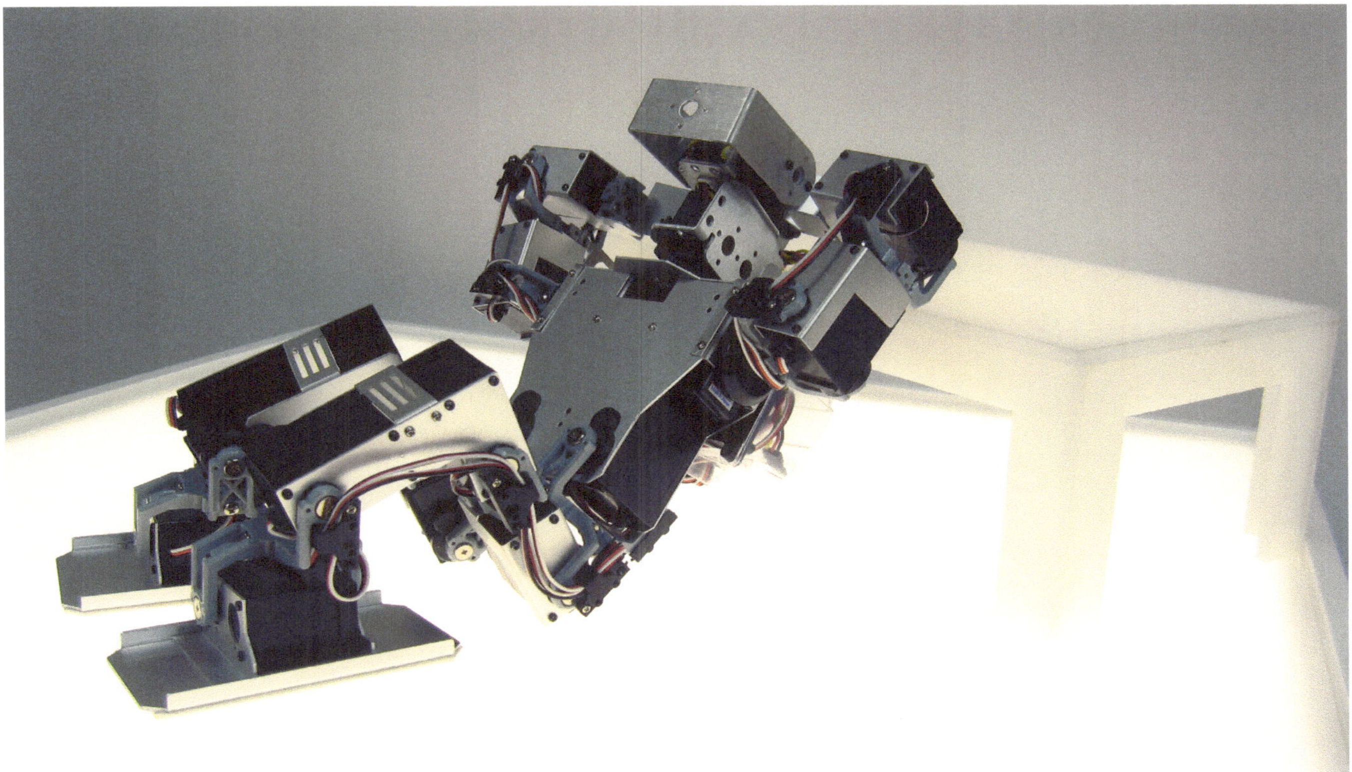

89.

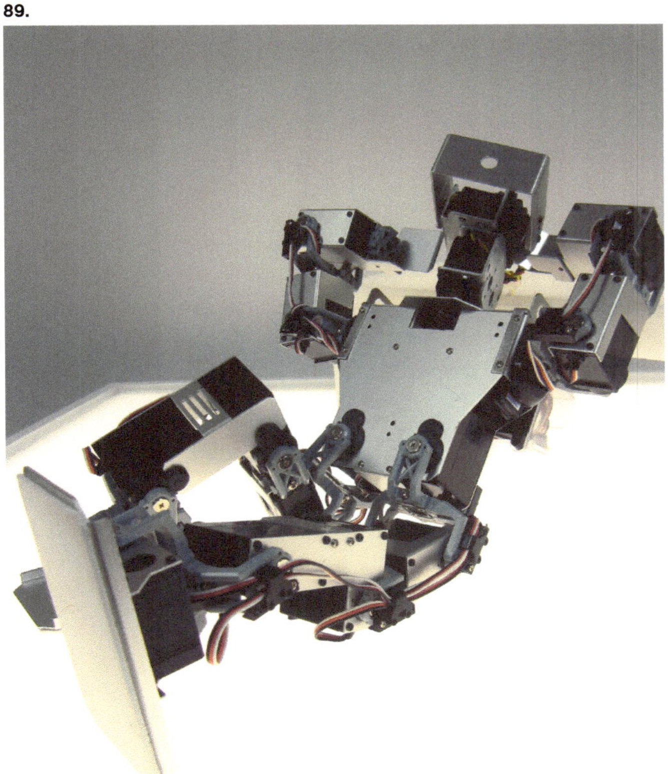

90.

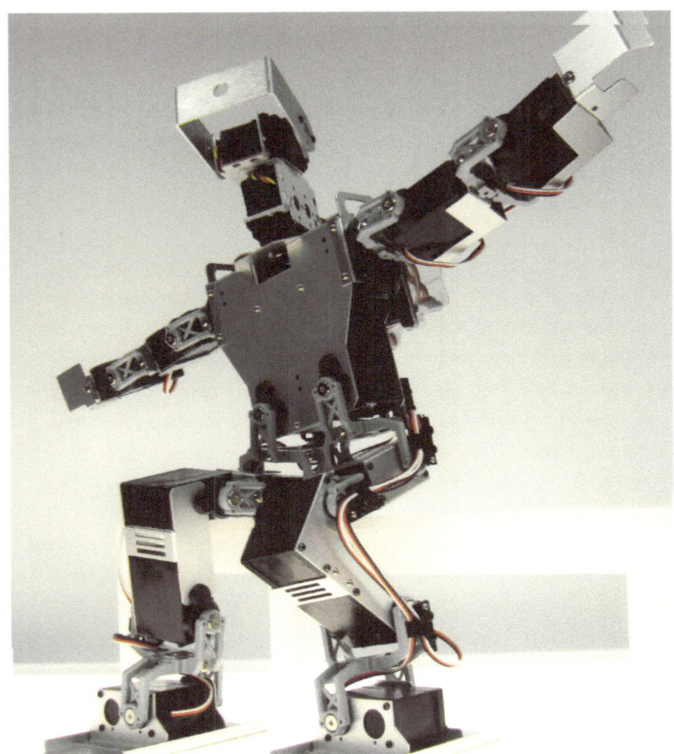

91.

89–91. Images from Fernando Orellana, *Sleep Waking*, 2008

SWAMP (DOUGLAS EASTERLY AND MATTHEW KENYON) + TIAGO RORKE
THE TARDIGOTCHI
2010

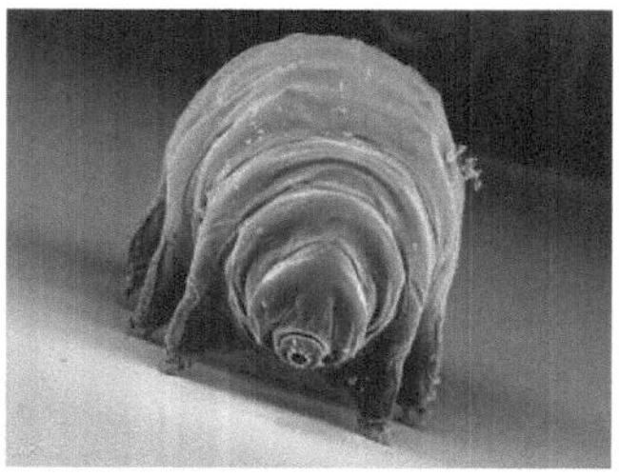
98.

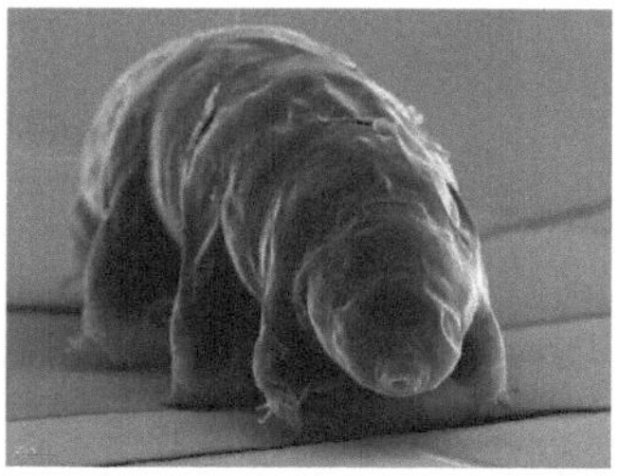
99.

98, 99. Microscopic tardigrades

As its portmanteau name suggests, the Tardigotchi[1] is a hybrid between two pets: a toy avatar derived from the 1990s Tamagotchi and a less famous but living organism called a tardigrade.

Tardigrades,[2] also known as "water bears" or "moss piglets," are microscopic animals with eight legs. They can be found all over the world and are able to survive in extreme environments. Not only can they endure both extremely high and low temperatures, but they also can tolerate 1,000 times more radiation than other living creatures and can live for almost a decade without water. They are the only animals known to be able to survive the vacuum of space.

Tardigotchi is a composite of the living tardigrade and its digital avatar. In this work the tardigrade lives on one side of a portable, brass-cased computing sphere; on the other, an LED screen displays a digital tardigrade. When you hold Tardigotchi, you simultaneously hold biological life and digital life—it's an animal, a pet, a game, and an interface.

Once a day *Tardigotchi* digitally signals the owner that it is hungry. The owner must then place it on a docking station, where a syringe can poke through the silicone wall of the tardigrade's home and pump in some moss water. Meanwhile, a microcontroller relays the feeding animation to the avatar. After the docking mechanism has removed the syringe from the miniature ecosystem and pulled the apparatus back into a neutral position, the avatar loops through a short animation that displays its full belly. Both the real and digital pets have been fed.

Tardigotchi is a pet you needn't ever worry about. You can email it and even contact it through Facebook. When you send a message to *Tardigotchi*, a Bluetooth signal conveys the information to the sphere, turning on a small incandescent lamp and warming the tardigrade's enclosure. At the same time, it also runs a short animation showing the avatar basking in the sun.

1. http://www.tardigotchi.com/
2. http://en.wikipedia.org/wiki/Tardigrade

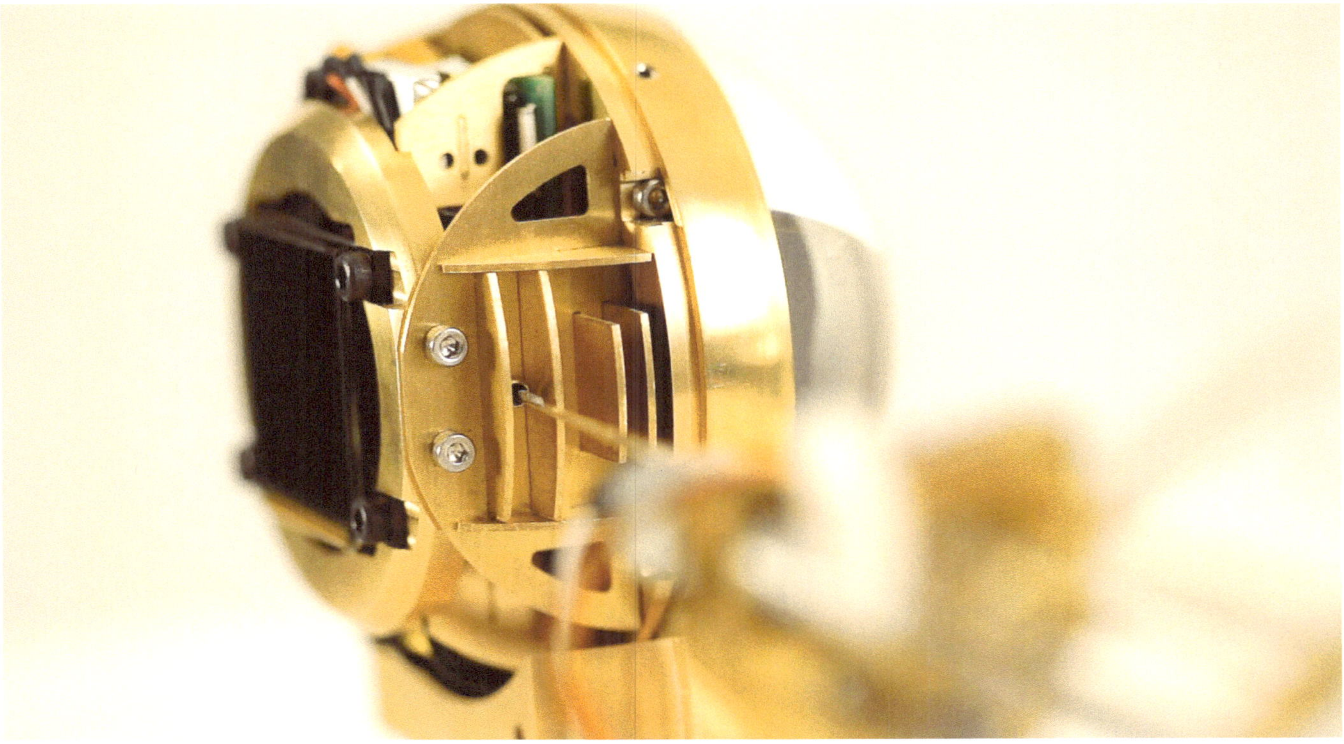

100.

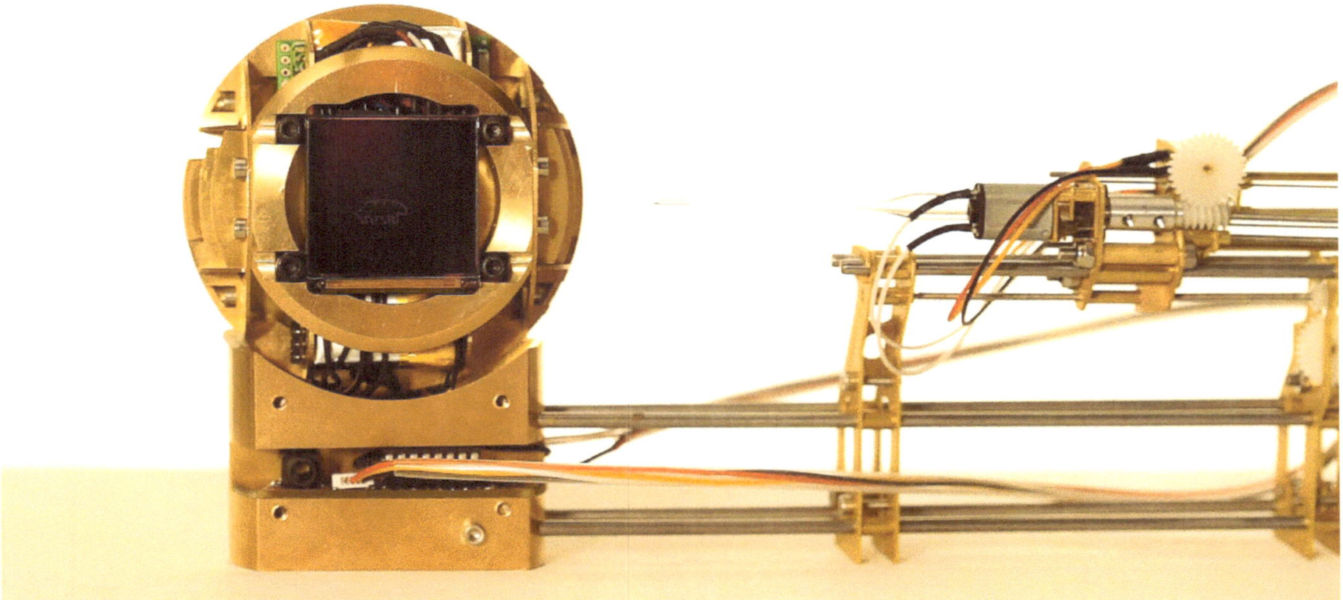

101.
100, 101. Images from SWAMP and Tiago Rorke, *Tardigotchi*, 2010

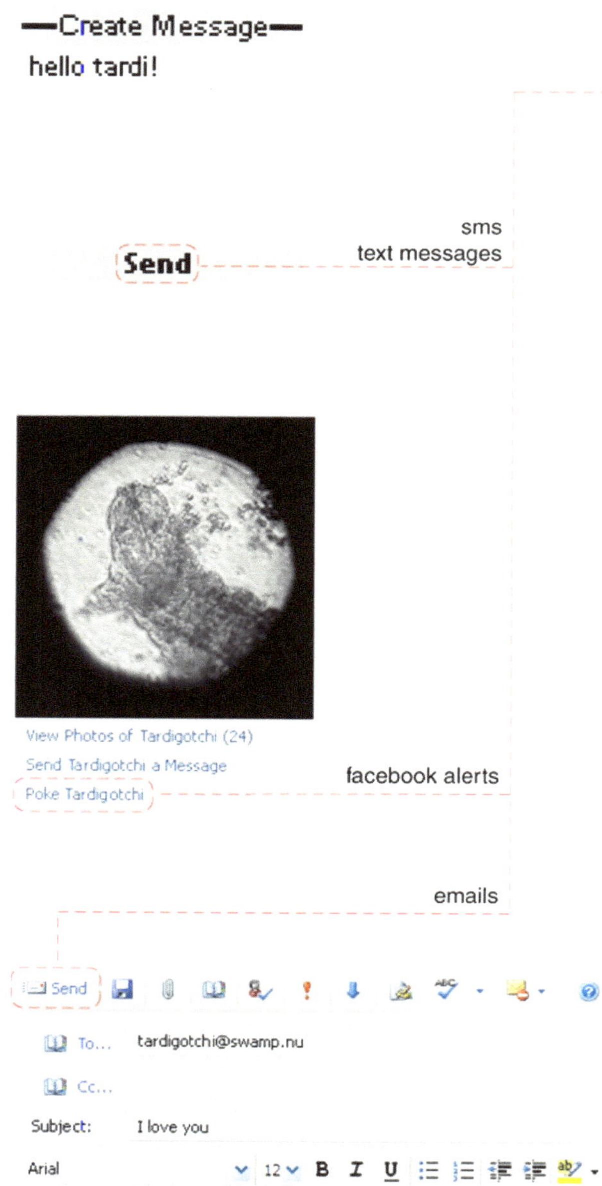

102. Page from the Tardigotchi website

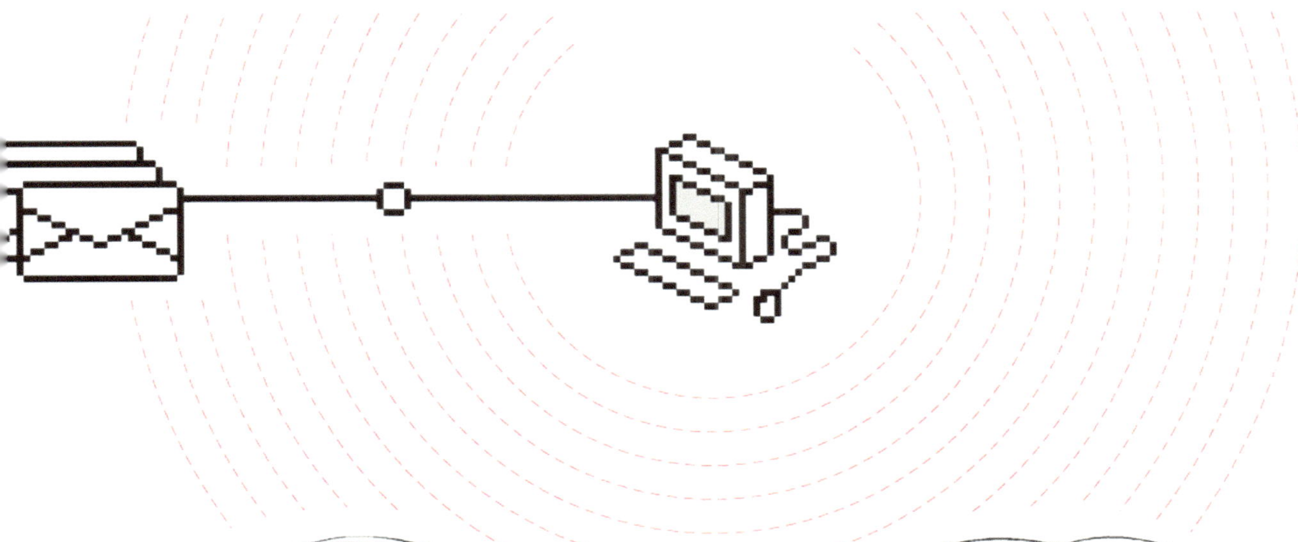

bluetooth connectivity

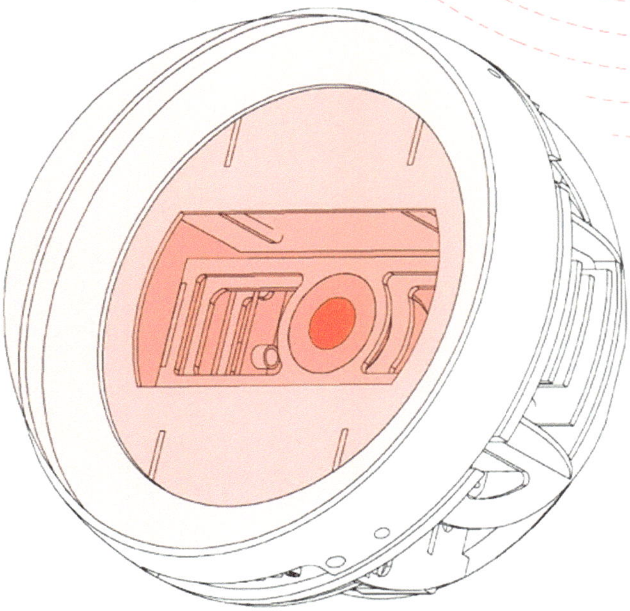

tungsten lamp illuminates the enclosure

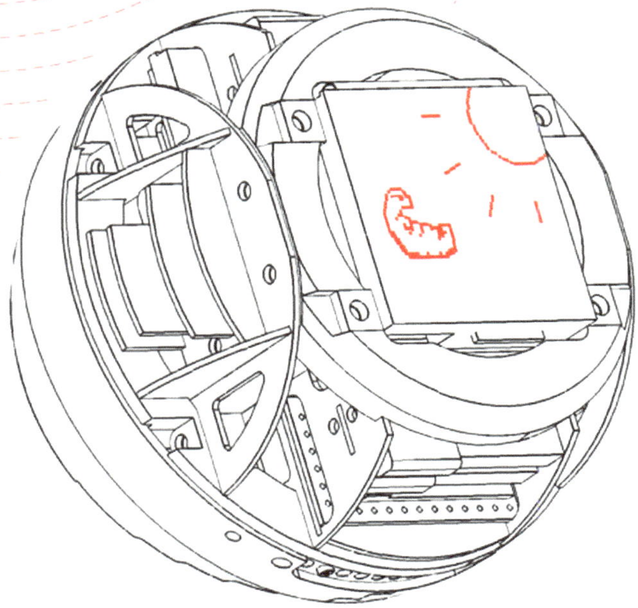

digital waterbear gets some sun

AGNES MEYER-BRANDIS
CLOUD CORE SCANNER
2010

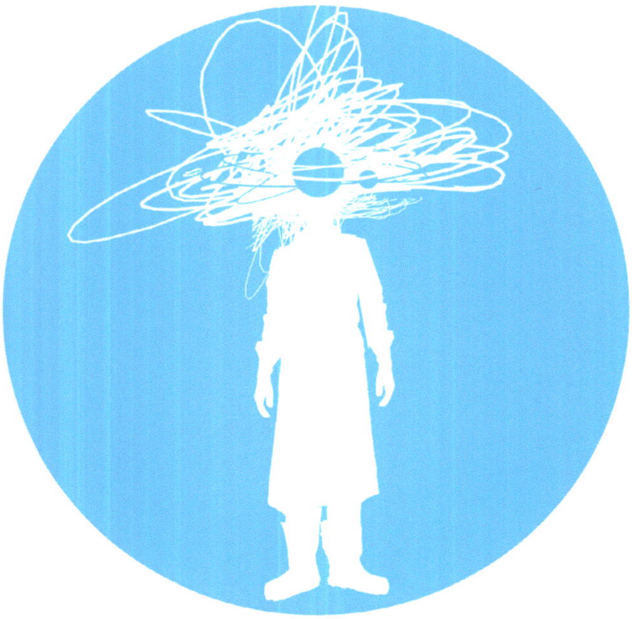

103. Agnes Meyer-Brandis, *Cloud Core Scanner Logo*, 2010

In September 2007, Agnes Meyer-Brandis was invited to participate in a DLR (German Aerospace Center) zero-gravity[1] flight. Although such flights are primarily reserved for scientific purposes, Meyer-Brandis used the opportunity to work on her art project *Cloud Core Scanner* under conditions of temporary weightlessness.

During parabolic flight, an aircraft lifts its nose toward the sky at an angle of forty-five degrees and is "pushed over the top" to achieve a parabola. At the peak of this curve, the crew experiences about twenty-five seconds of weightlessness.

Meyer-Brandis participated in the parabolic flight to study what she calls "cloud cores" in a state of weightlessness. Small particles suspended in the air form the basis for clouds. The water in the air condenses around these particles, or cores; this accumulation of water becomes a drop; and many drops generate a cloud. The artist has developed probes capable of observing these otherwise elusive (i.e., imaginary) cloud cores.

Meyer-Brandis took two instruments on board with her. The first one, the *Cloud Core Scanner*, is the first ever picture-generating instrument for examining the behavior of particles in the cloud cores during the zero-gravity flight. The second apparatus is the ADM-Filmbox, which she used to capture images from even the most fleeting cloud cores. The cloud core world can be observed inside what looks like little snow globes.

1. http://en.wikipedia.org/wiki/Weightlessness

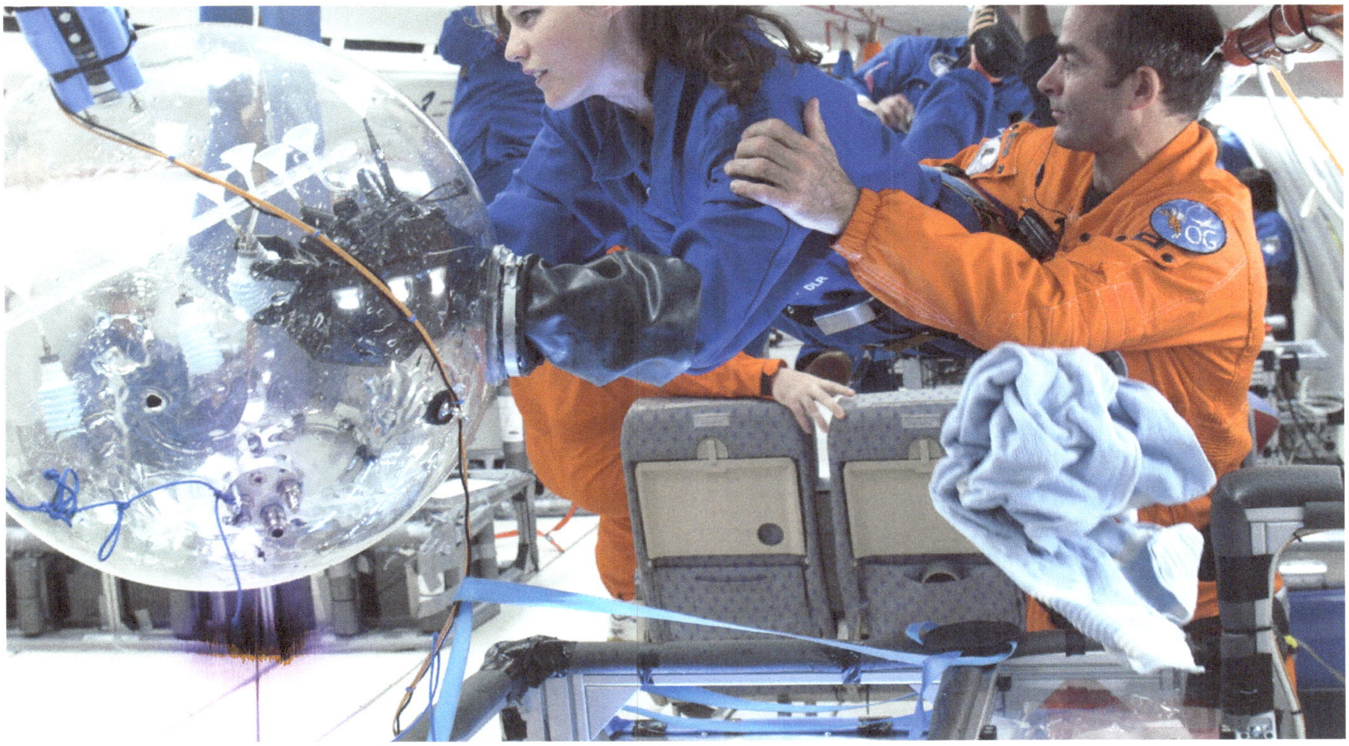

104. Agnes Meyer-Brandis at work during zero-gravity flight

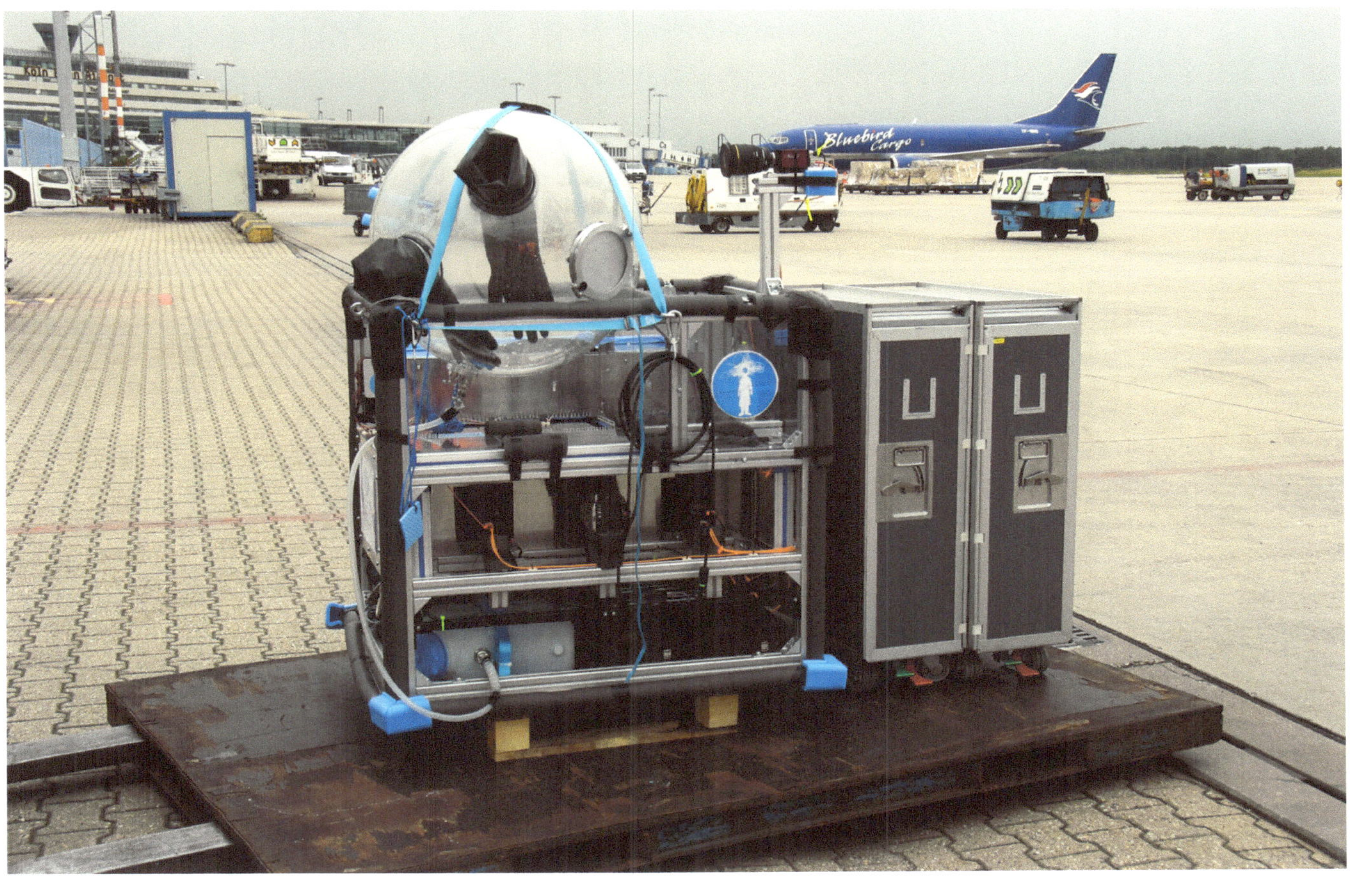

105. Embarking the *Cloud Core Scanner* on the zero-gravity flight

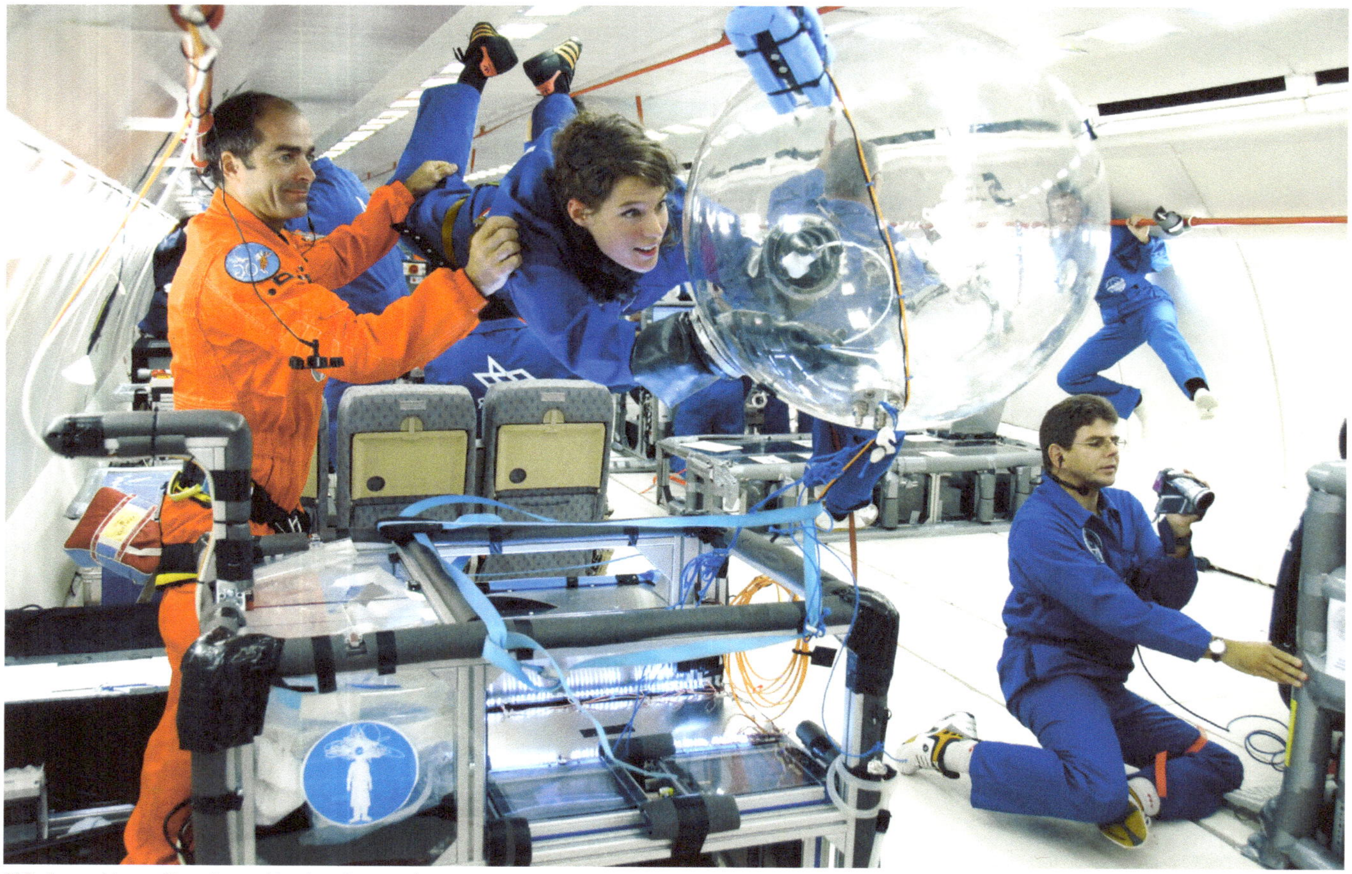

106. Agnes Meyer-Brandis working in microgravity

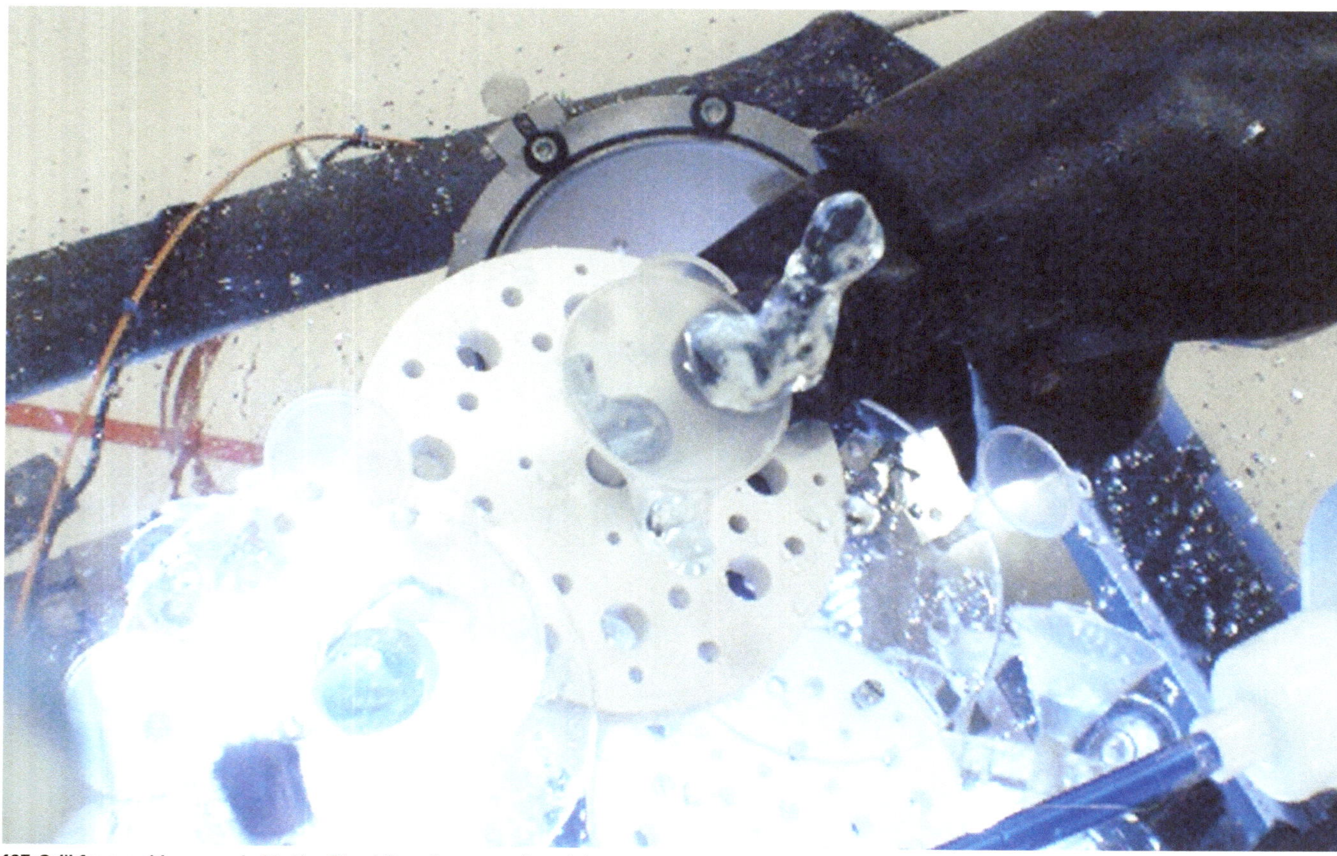

107. Still from a video recorded in the *Cloud Core Scanner* sphere lab

Adam Zaretsky

As an artist, how do you see your role in a technological or scientific setting?

As an artist, my role in a scientific or technological environment is unsutured, torn and also cozy. I often feel in a double bind as the funding sources expect my work to aid in the social acceptance of new technology or provide a soft social debate without any influence on the pace and direction of future research.

If I speak fondly of scientific processes as invasive, meddlesome, sadistic methodologies, I am working for the power structure, the tribe of devilish successfuls, who doom the earth with their engineering attitude and their inability to perceive thanatos in their optimization dreams.

On the other hand, if I speak from the scientific environment as a critic-antagonist, I am perceived of as a corporate pseudo-luddite, lackey muckraker for hire. What kind of hypocrite speaks on a disdainful bully pulpit from the lab and payroll of innovation, national, scientific and corporate funders?

Finally, if I make a mess that is off sauntering in a transgenic ecocidal swath of destruction, it is the fault of the non-professional, the artist, the buffoon and this seals the insurance of power's resilient use of the arts to interpolate technoscientific introduction into the ecosphere and markets, markets, markets.

My self appointed role is always to try to communicate an immersed and informed, 'demystified' version of the libido which drives the obscure processes of scientific research to the public and scientists themselves. Meanwhile, I continue engaging in the mystical practices of jazz parrahesia (screaming-angry-joyous-postlogical-ranting) through art-sci sex, blood and politics with a fidelity to voice-acts-cross/species_kindred, unencumbered by expectation.

I'm at home with the sexual frenzy and the blood dripping politics of: the world, the lab, the bedroom, and the art studio.

So the confusion and anger I add to the orgy of dismantling in these realms is the cozy role I play in this arena.

SYMBIOTICA/TISSUE CULTURE & ART PROJECT
VARIOUS PROJECTS
2000-ONGOING

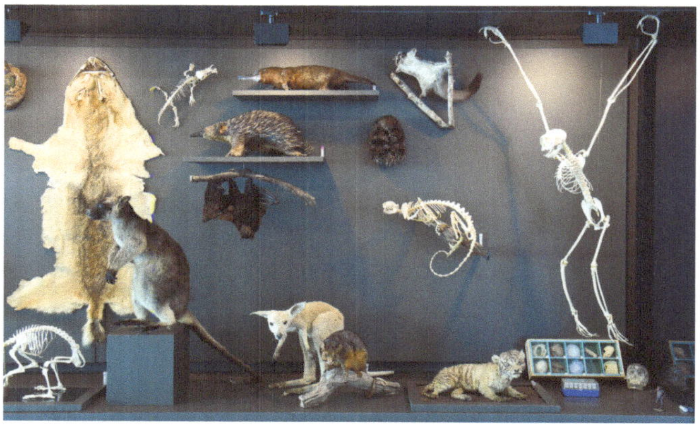

108. ***NoArk I***, **The Tissue Culture & Art (Oron Catts & Ionat Zurr), Vessel design in collaboration with Marcus Canning, taxidermy, preserved specimens, bioreactor, glass bids and living cells. 2007**

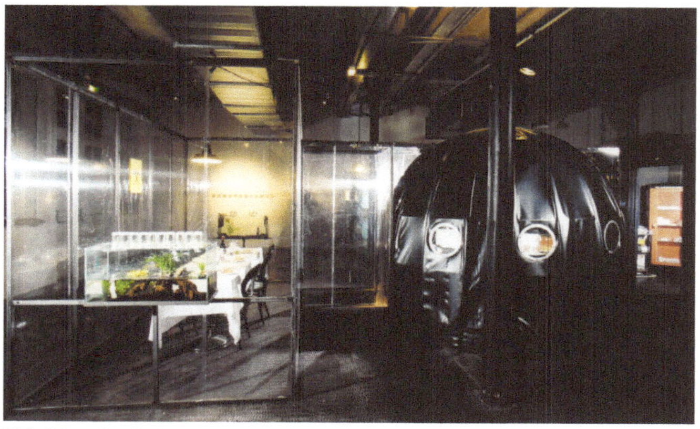

109. ***Disembodied Cuisine*** **Installation Nantes France 2003, The Tissue Culture & Art Project, 2003. Photography: Axel Heise**

Before establishing SymbioticA[1] in 2000, Oron Catts founded, together with Dr. Ionat Zurr, the Tissue Culture and Art Project[2] to explore tissue culture manipulation and engineering as a form of inquiry into shifting perceptions of what constitutes life.

The group gained worldwide recognition in 2003 with *Disembodied Cuisine*,[3] an installation and performance that involved the growth, cooking and public tasting of tissue-engineered meat from a frog that was kept alive during the whole event. As curator and writer Jens Hauser[4] notes, the project contributed to the open use of existing knowledge toward political ends: by bringing the concept into the public domain, the artists made it difficult for commercial firms to patent and make a profit out of "tissue-engineered meat."[5]

Perhaps the group's most iconic work, *Victimless Leather*[6] saw the creation and development of a semi-living garment grown out of "immortal" cell lines. "Biotechnology is producing more and more chimeras," explains Catts.[7] "Human stem cells are implanted into rat brains where they develop into entirely functional nerve cells. Biotechnology breaks down the barriers between species, and this triggers a series of questions: how many human cells does a rat brain need before it becomes human? And how many animal organs can we implant into human bodies before they become [nonhuman] animals themselves?"

More recently, the *NoArk*[8] project directed attention to the taxonomical crisis caused by the arrival of new, hybrid life forms engineered through biotechnology. In *NoArk*, Tissue Culture and Art Project displayed a series of species inside a modern version of a Renaissance cabinet of curiosities. Some of these species were familiar and presented in taxidermied form, such as a crow, a rat, and the head of another mammal. The others, however, were not only unknown to most people, but also alive in the display. They comprised a collection of cells and tissues engineered from different species, compiled from tissue banks, laboratories, museums, and other collections. Although these engineered suborganisms are alive, natural history museums don't take them into account. They remain hidden in laboratories, floating in the limbo of unclassifiable biotechnology.

Catts's individual ongoing project *The Autotroph*[9] extends his practice to address ecological concerns and global environmental pressures. *The Autotroph* is a kinetic sculpture that functions as a desalination plant; it attempts to protect Lake Clifton in Western Australia (or at least a portion of it) from the effects of climate change and urban development. The ultimate goal is to contribute to the conservation of a population of thrombolites,[10] rock-like structures built by microorganisms and regarded as one of the earliest forms of life on Earth.

1. http://www.symbiotica.uwa.edu.au/
2. http://www.tca.uwa.edu.au/
3. http://www.tca.uwa.edu.au/disembodied/dis.html
4. http://archive.transmediale.de/page/detail/detail.0.persons.648.3.html
5. Jens Hauser, "Bio Art—Taxonomy of an Etymological Monster," presentation at Ars Electronica 2005.
6. http://www.tca.uwa.edu.au/vl/vl.html
7. http://www.arte.tv/fr/Art-biotech/796180.html
8. http://www.tca.uwa.edu.au/noark.html
9. http://symbiotica-adaptation.com/?page_id=44 and http://wiki.mech.uwa.edu.au/index.php/The_AUTOTROPH_-_De-Salination_Plant
10. http://en.wikipedia.org/wiki/Thrombolite

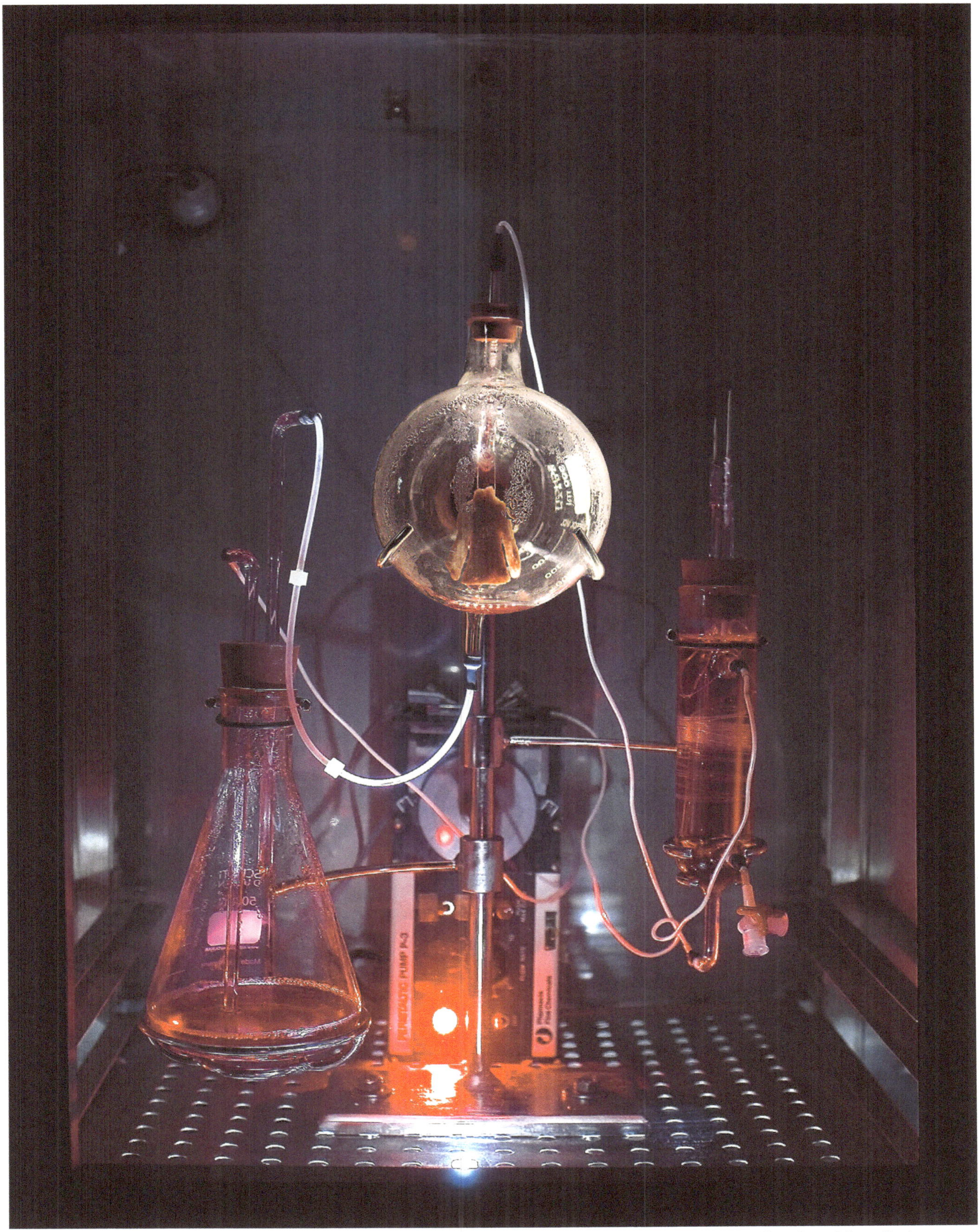

110. ***Victimless Leather- A Prototype of Stitch-less Jacket grown in a Technoscientific "Body"***, The Tissue Culture & Art (Oron Catts & Ionat Zurr), Biodegradable polymer connective and bone cells, Dimensions variable, 2004

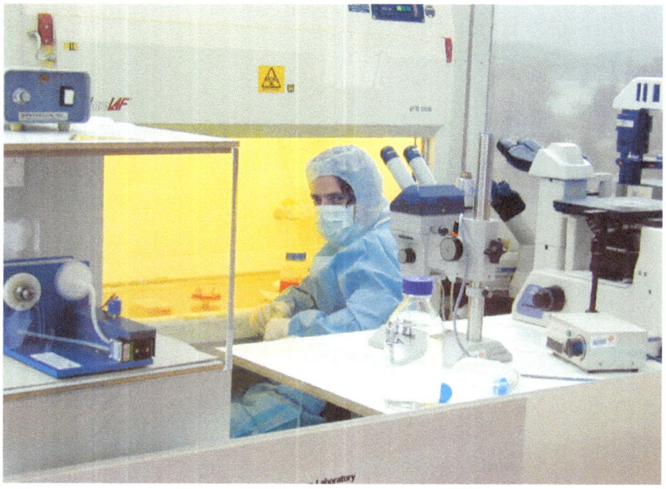

111. SymbioticA and Arts Catalyst Workshop, Kings College in London

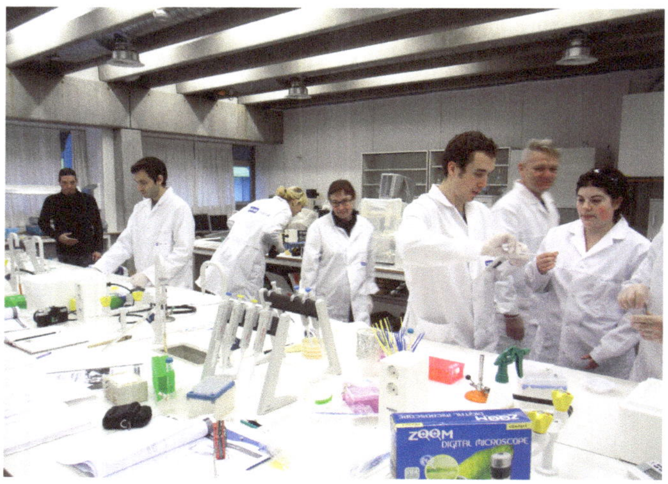

112. 'SymbioticA Biological Arts Masterclass' Stavanger, Norway, 2008, photo credit: James King

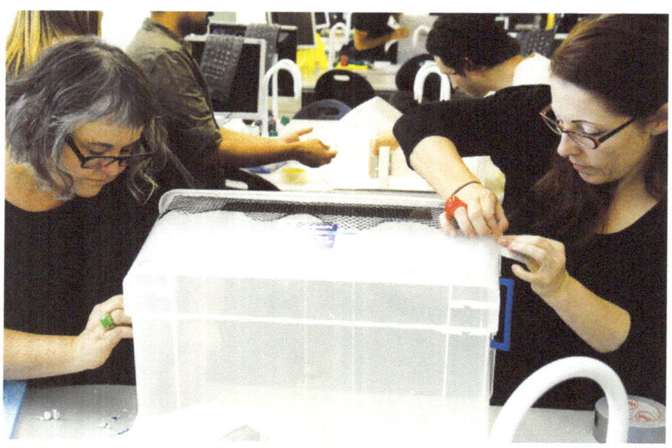

113. SymbioticA BioTech Art Workshop, RMIT University, Melbourne, Australia, Photographer: Pete Waters, November 2009

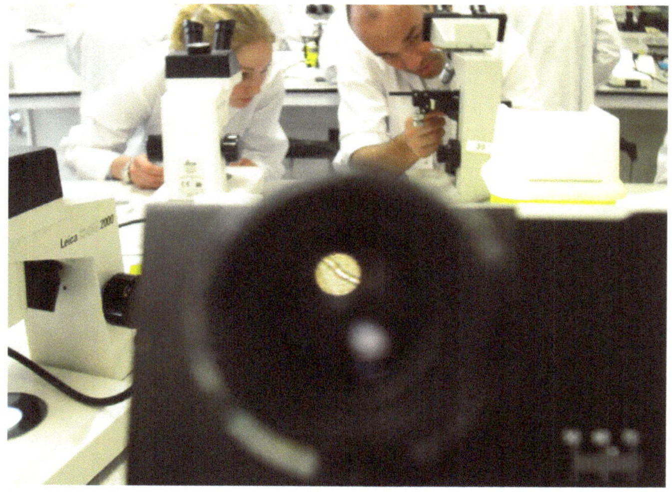

114. SymbioticA and Arts Catalyst Workshop, Kings College in London, 28 March - 2 April 2005

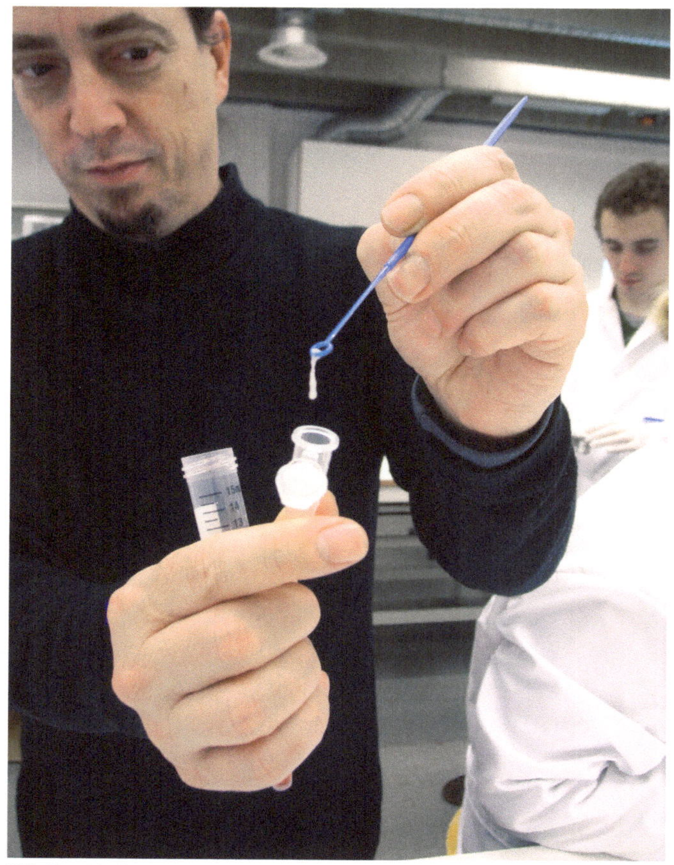

115. 'SymbioticA Biological Arts Masterclass' Stavanger, Norway, 2008, photo credit: James King

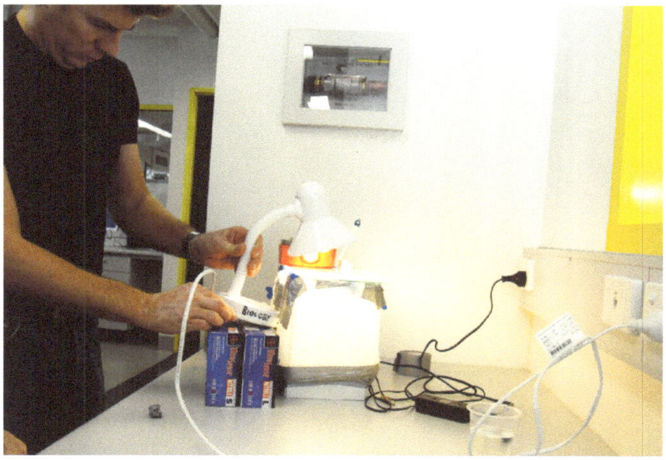

116. SymbioticA BioTech Art Workshop, RMIT University, Melbourne, Australia, Photographer: Pete Waters, November 2009

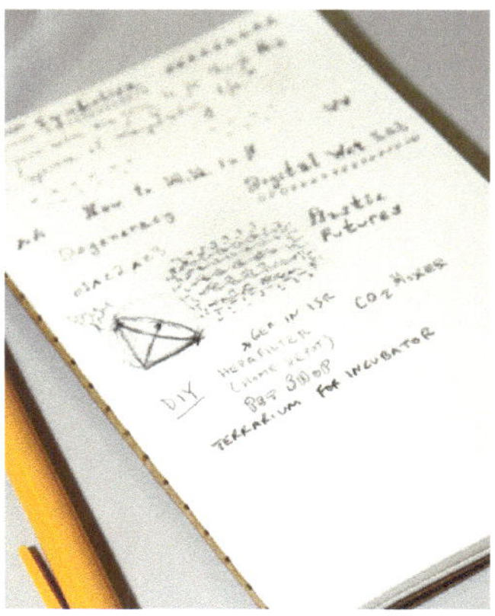

117.

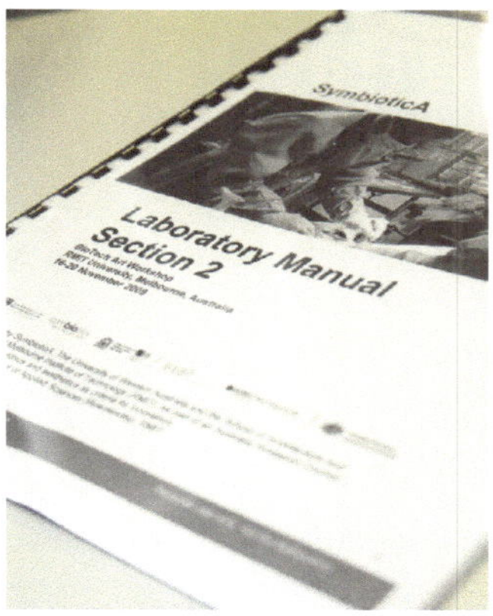

118.

117, 118. SymbioticA BioTech Art Workshop, RMIT University, Melbourne, Australia, Photographer: Pete Waters, November 2009

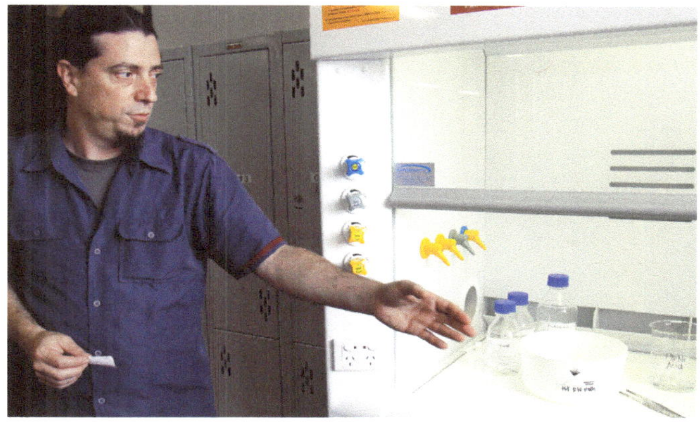

119. SymbioticA BioTech Art Workshop, RMIT University, Melbourne, Australia, Photographer: Pete Waters, November 2009

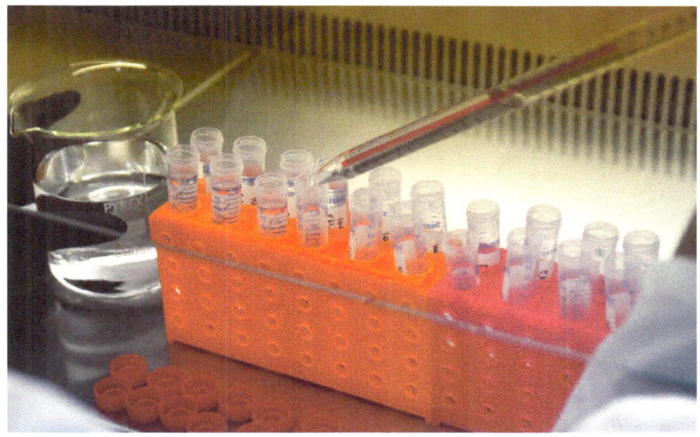

120. SymbioticA BioTech Art Workshop, RMIT University, Melbourne, Australia, Photographer: Pete Waters, November 2009

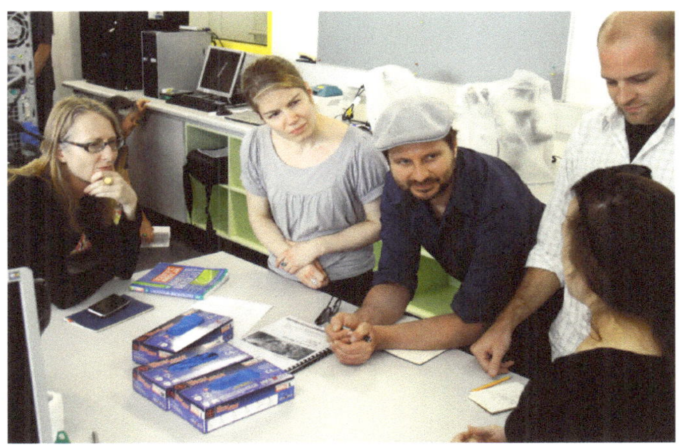

121. SymbioticA BioTech Art Workshop, RMIT University, Melbourne, Australia, Photographer: Pete Waters, November 2009

Geraldine Juarez

As an artist, how do you see your role in a technological or scientific setting?

The thing is... technology is all around, for me it is not a specific context, I think everyone is immersed in this time in a technological environment.

With FAT, technology is our tool, our theme and our means of production and distribution of our work. The role of FAT is mostly to turn mainstream all the open source culture that sometimes is seen as something nerdy or inaccessible, when really it is not. Our role is to be the Tootsie Pop of the pop culture :)

With telecomix, it is different. We are actually more like social technicians who use technology to maintain technology open. My role in telecomix is not as artist though, although the whole assemblage of telecomix sometimes is a work of art. We produce videos, infrastructures, writting and theory to make cool the duty of protecting the internet and mostly keep open any channel of communication.

I guess in both projects I don't see my self as someone with a specific role... but as part of a group. With FAT our role is to develop cool open source and sharing culture. With telecomix the role is to protect internet and create political awareness, plus provide emergency communication when needed. In telecomix some people are very scientific, such as real engineers and skilled programmers, they don't talk about their role in science, they just do science.

THE MAKER MOMENT

MAKING, TINKERING, DIY, DIWO, HANDMADE, FUNCTIONAL, RECYCLED, REPURPOSED, HOW-TO, CRAFT

In the wake of the Soviet Union's launch of Sputnik in 1957, the United States attempted to close the perceived gap between Soviet and U.S. scientists by establishing the National Defense Education Act (NDEA) of 1958, which provided $887 million for improvements in education over four years. Among the Act's areas of focus were science, mathematics, and technical fields. The years that followed saw the expansion of "shop" classes in schools, the institution of classroom labs, the proliferation of science fairs and math clubs, and the creation of award incentives like the Westinghouse Science Honors Institute for high school students. Through the NDEA, hundreds of schools received funding for the construction of planetariums and space science classrooms within their buildings. NDEA was a top-down directive: the idea was to create a new generation of scientists and engineers to help the United States maintain its military status and competitiveness in a global economy.

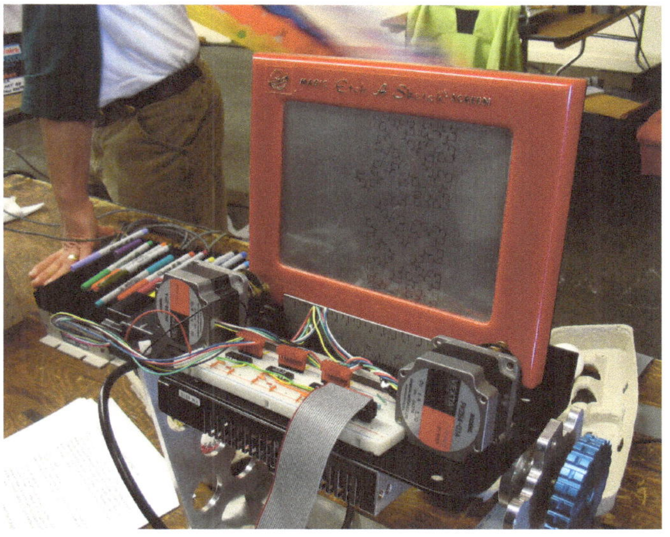

122. Bruce Shapiro's programmable etch a sketch at Maker Faire 2006

The type of training that many baby boomers received during this period has faded from the agenda of American public schools today—to say nothing of the fact that creativity receives even less of a focus. As many have pointed out, the U.S. government's more recent STEM (Science, Technology,

Engineering, and Mathematics) education focus lacks STEAM because of its omission of Art—a key ingredient in the alloy of innovation.

Enter the Maker Movement: a grassroots, peer-to-peer, open source cultural force that embodies a hands-on approach to building, modifying, and repairing things. Popularized by the publication of *Make* magazine, which appeared on newsstands in 2005 and has since served as a principal organizing voice, the Maker Movement has empowered countless basement tinkerers to seek community and share their projects with one another. Makers are no longer working alone in this joyful tampering and building. Hacker spaces around the world now provide a kind of shared "studio space," where for a modest membership fee, one can gain access to a basic machine shop, pooled materials, and a diverse knowledge network. International associations like Dorkbot have also created a meeting place for artists, engineers, designers, scientists, and inventors to present their ideas and processes in plain language.

An alternative economy has emerged around maker culture, one positioned outside the mainstream marketplace. Thanks to Kickstarter.com, a site launched in 2009, makers can now raise funds for their projects through "crowd-funding." Sites like Etsy.com also serve as an alternative marketplace for selling handmade products and inventions.

The maker methodology thus inverts the former NDEA model of top-down education and replaces it with a grassroots network of DIYers (Do It Yourself-ers) and DIWOers (Do It With Others-ers) who revel in complex projects adapting antiquated electronics, household materials, and modified consumer products. A maker slogan says it all: "If you can't open it, you don't own it."

In recent years, the institutional science world has begun to take notice. In 2010 the National Science Foundation made an appearance at the World Maker Faire in New York, and it has since sponsored related convenings, like the 2011 "Art as a Way of Knowing" conference at The Exploratorium in San Francisco. Last year Congressman Bill Foster of Illinois introduced the National FabLab Network Act of 2010, providing for "the establishment of... community-based, networked Fabrication Laboratories across the United States."

The Maker Movement champions the handmade and the functional, the incorporation of everyday materials, and the inclusion of the untrained and nonprofessionals in the creative process. In general, it exhibits a reluctance to participate in the mainstream culture of mass production and an opposition to proprietary technology—makers share information about hacking and modifying commercially manufactured products. In this way, the Maker Movement has origins all over the creative sphere: in the Arts and Crafts movement, in folk art, in industrial design, and in punk rock.

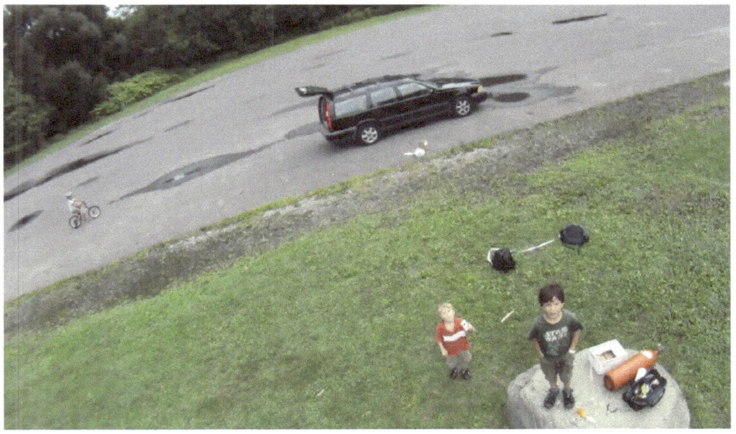

123. Video still from amateur space footage shot by Max and Luke Geissbuhler

Artists have long embodied the values of maker culture: to use the materials immediately available, to recycle, to make and grow one's own, to create new forms from old ones, to customize, and to approach materials with curiosity and a desire to understand the way things work. In a sense, maker culture is the popular adoption of artistic values that have existed at least as far back as the fifteenth century, a time when Renaissance humanism asserted that individuals had the capacity to embrace, understand, and apply all areas of knowledge. The difference is that while artists once made things from nothing, makers and their allies make things from other things. And the end product can have, literally, astronomical implications.

In August 2010, Brooklyn dad Luke Geissbuhler and his son Max managed to send a homemade spacecraft nineteen miles into the sky to record video from the Earth's stratosphere. The duo took eight months to build their craft, which included an iPhone, GPS equipment, an HD video camera, and some hand-warmers, all housed in an insulated case and pulled by a nineteen-inch weather balloon. A seven-year-old child sending a homespun craft into space: this is something that if stated in the mission of the 1958 National Defense Education Act, would have seemed spurious and absurd.

And yet it is somehow its consequence.

James L. Acord

(1944–2011)

James L. Acord (1944-2011) was an artist who worked primarily with radioactive materials. He was the only private individual in the world licensed to work with radioactive material. The world's first nuclear sculptor spent years learning how to execute modern alchemy: the conversion of radioactive waste into inert material and subsequently into sculptures. He spoke on art and nuclear science at both art and nuclear industry events in the US and the UK and organized many forums that brought together artists, activists and nuclear industry experts.

The development of his artistic process, from the simple discovery that granite is mildly radioactive, to moving to live on the outskirts of the Hanford Nuclear Reservation, the most contaminated nuclear site in the United States, where Plutonium was first isolated, was part of a 20-year-long living performance which attempted to create sculpture and events that probed the history of nuclear engineering and the environmental implications of long-term storage of nuclear waste. His major ambition while there was to build a "nuclear Stonehenge" on a heavily contaminated area of land in the site, incorporating twelve uranium breeder-blanket assemblies.

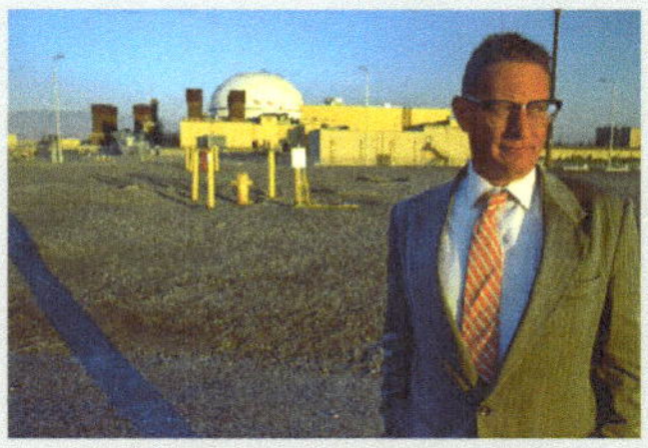

124. The nuclear sculptor James L. Acord stands in front of the Fast Flux Test Facility, Hanford Nuclear Reservation, WA, USA. James Acord archive

MACHINE PROJECT
VARIOUS PROJECTS
2004-ONGOING

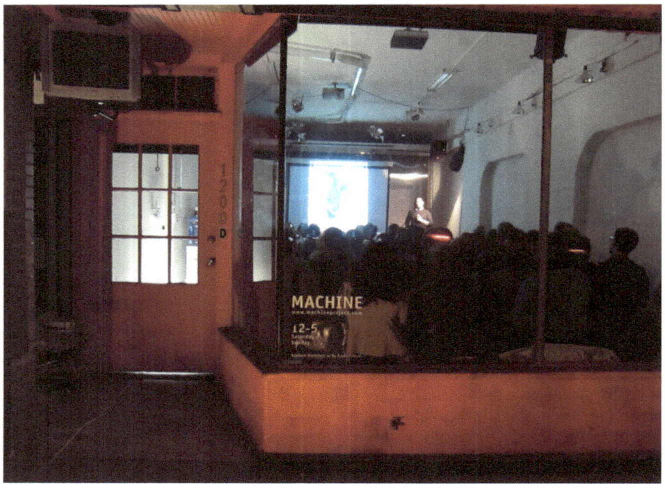

125. Machine Project's Los Angeles storefront, 2011

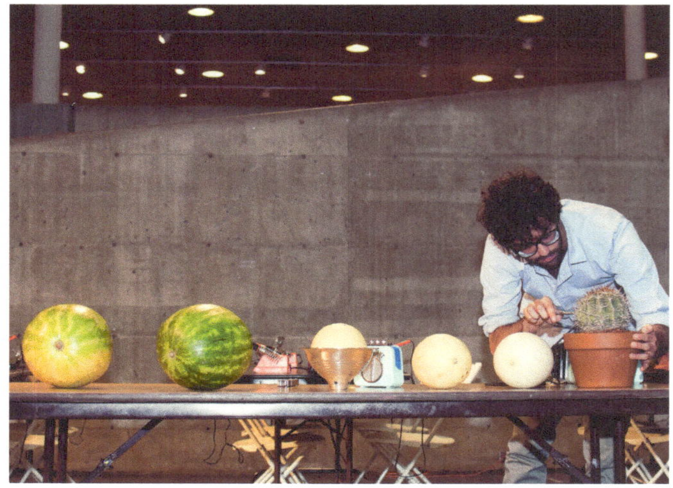

126. Fallen Fruit's Electronic Melon Drum Circle workshop at LACMA, 2010

Machine Project is located in a storefront in the Echo Park neighborhood of Los Angeles. Begun in 2004 by artist Mark Allen, Machine is a "not-for-profit arts organization and community event space dedicated to making specialized knowledge and technology accessible to artists and the general public." On any given day at Machine, one might take a workshop on "Basic Electronics for Artists," "Intro to MIG Welding," or "Programming the iPhone." Acting as a platform for artists to present a huge number of live events, performances, installations, poetry readings, and other more uncategorizable events, Machine has developed a reputation for integrating unlikely subjects in a single evening: "Entomology Meets Etymology," for example, or "Lay Science Puppet Performances."

Machine describes its terrain as encompassing "art, technology, natural history, science, music, literature, and food," as well as scores of other keywords that are listed on their website. Machine's style of presenting promotes intellectual whimsy and hands-on engagement that makes rarefied knowledge accessible. The storefront location is a strategic vehicle for these transactions; it is intentionally in a high traffic (not hard to find) area, and the shop's front window invites curiosity and entry. Outside of the storefront, Machine operates as a loose confederacy of artists producing shows at locations ranging from the Santa Monica beach to the Los Angeles County Museum of Art and the Hammer Museum.

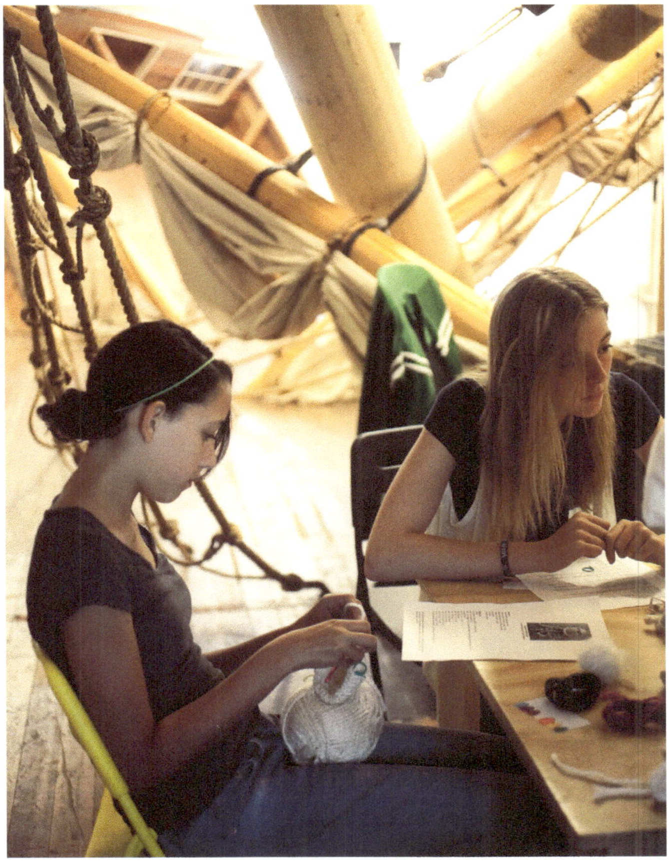

127. Crochet workshop aboard Josh Beckman's Sea Nymph.
Photo: Marianne Williams, 2010

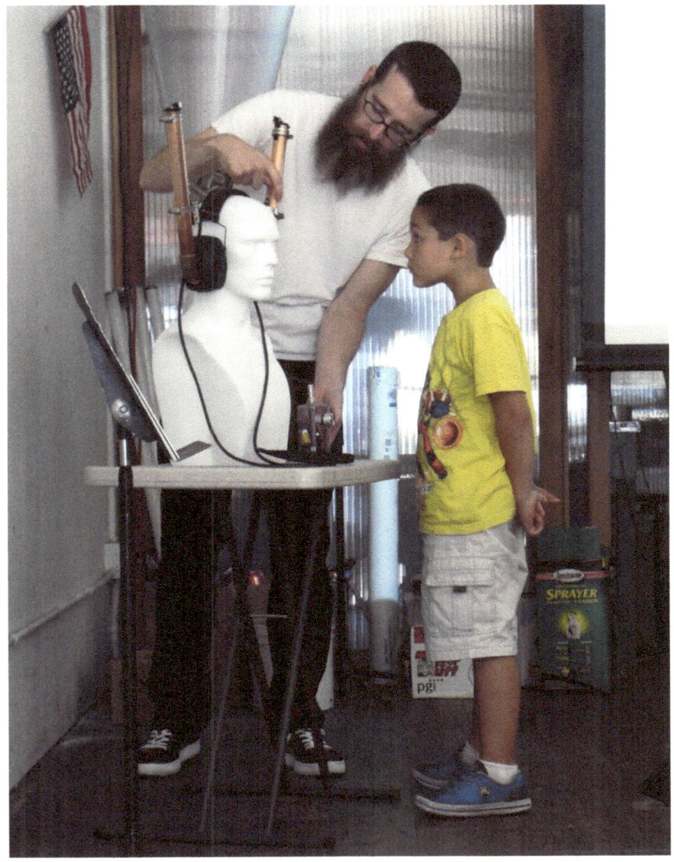

129. Alex Braidwood's listening experiment, "Noisolation Headphones"
at Machine Project, 2011

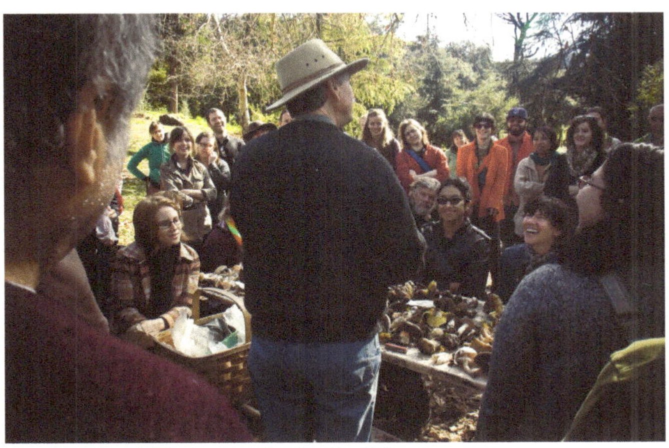

128. Fungifest field trip, organized by David Fenster.
Photo: David Fenster, 2010

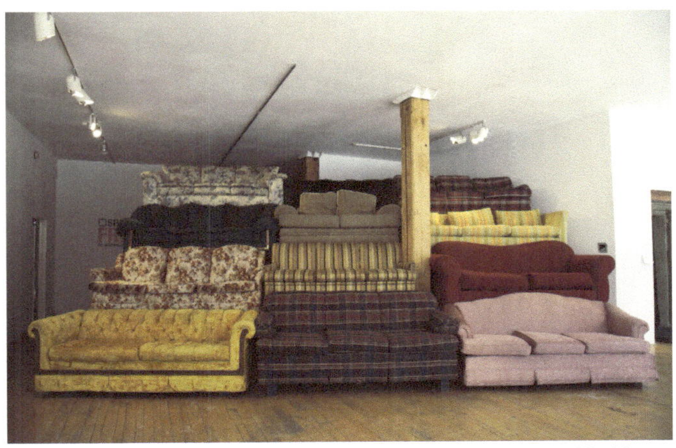

130. Nate Page's couchbleachers made from thrifted couches for Machine
Project's exhibition at Spaces, Cleveland, OH, 2011

Skype interview

Mark Allen

Founder of Machine Project, Los Angeles
Sunday, February 20, 2011

ANDREA GROVER: Thanks for putting aside some time for me. Well, we have essentially 170 pages right now, and it's looking like a compendium of examples of artist works from the last five years that touch upon art, science, and technology. But then there are also chapters that contextualize the work within a set of methods. And we have this crazy chronology that telescopes back to 25,000 BC and includes everything from when the first Kinko's opened to the launch of Kickstarter.

MARK ALLEN: I just learned that the rotation of the earth is slowing down.

AG: Yeah, I heard that too, and there was some specific reason for it, right? Something we did?

MA: No, it's been slowing down since the beginning. That's what happens: planets start slowing down after a while. Like three and a half billion years ago, the day/night cycle was only twenty hours long. So the days are actually getting longer.

AG: So we've added four hours? Days are getting longer? Goddammit!

MA: And eventually one side of the Earth will be night and one side will be day.

AG: Are you all getting this, everyone? OK, thank you for that. So, I have a set of five questions. We're trying to figure out what are the contributing factors to this moment in art, science, and technology work.

MA: OK.

AG: What was your impetus for starting Machine?

MA: I was particularly interested in making a center where different kinds of cultural discourses could intersect with each other. So, rather than segmenting it into a space for "contemporary art," "poetry," or "science," I was thinking about how the social [whole] could facilitate a transdisciplinary discourse around different communities of interest.

AG: So Machine "makes specialized knowledge and technology accessible to artists and the general public." I think there is something very specific about the way that Machine does that—in a both playful and whimsical way, but also by presenting large volumes of material.

MA: Yeah, of course. I think that part of our methodology involves figuring out how to make information—which might be esoteric, dense, or intimidating —accessible without diluting the content. And so a lot of our methods include

different kinds of social mechanisms: making things friendly, making things funny, making things social.

AG: Can you maybe give us some examples of emblematic Machine events that did that? Or continue to do that?

MA: Rather than one event, it's more about how they relate to each other in a continuum. For example, in one week we have a poetry reading, and the next week we have a fry-b-que, and the week after that we have a lecture on the sex life of sea slugs. It's kind of like taking things that are more pop culture and things that are more academic, and presenting them with the same attitude. This creates a certain kind of editorial position on how culture is consumed. And then we're not afraid to use hooks or gimmicks—like we just did a car theft class for kids and parents together, which is a basic introduction to electricity and mechanisms, but it uses this very kind of over-the-top structure to do it.

AG: And kind of illegal.

MA: Well, it's not illegal to break into your own car.

AG: But "car theft" is the title, right?

MA: Well, yes, that's the title.

[laughter]

AG: OK. So something like the sex life of sea slugs—who would present that?

MA: That was by Patrick Krug. He's an academic who studies sea slugs and their sex lives. A lot of this is just meeting people who are good public speakers, and who are comfortable talking in a less academic environment. Of course, we had Peter's talk with a performance by Fol Chen, whose band had written a song about the sex life of sea slugs.

AG: At this kind of talk, what is the audience composition?

MA: It's really hard to say. Half of our audience is people who come because it's Machine, and the other half comes specifically out of community interest. You get some marine biologists, and then you get some people who are just curious about interesting things in the world. I think there's been a shift in how people consume information – [now it's] as entertainment. Like while they're supposed to be working: twenty years ago when you're killing time at work, you might be calling people on the phone or playing golf in your office; now you just look at random things on Wikipedia. So people now consume esoteric information as entertainment, and Machine is kind of the physical manifestation of this. I think of it as a blog made physical in a certain way.

AG: It seems that over the years there have been more people taking ownership of the space. I originally thought of you as the curator, but now I see more artists initiating all manner of projects at Machine, as if they know that it's fertile ground for these kinds of experimental approaches.

MA: Yeah, I started it as a venue, and then started inviting people to do projects, and it progressed into a collective activity. And then once we started doing things off-site—whether at the beach, a forest, or a museum—we began moving through spaces more as a collective entity than as a venue.

AG: I was just looking at your site, and the most recent activity happening at Machine was "Lay Science Puppetry" performances?

MA: Uh huh. That's a puppet show by Eric Lindley and Katie Shook, that's an experiment in how visual perception works. We're also doing something at the Berkeley Art Museum, which is called the "Confusatron." It's a workshop that [pulls together] four things we've presented before at Machine for four different kinds of audiences. For example, Tranimal (which is a style of drag makeup), plant cloning, kimchi making, and amplifying watermelons with contact microphones. It's sort of like a Machine buffet.

AG: A machine buffet.

MA: So you might get your drag makeup done, and then make some kimchi; or you might learn how to do a plant cloning, and then make an amplified watermelon; or you might do all four throughout the course of the night. So some of it is just wanting to present workshops that we think are interesting, but also to engineer a kind of collision of audiences.

AG: The amplified watermelon was originally used for a drum circle? Is that right?

MA: Yeah. We did that at LACMA for a project with Fallen Fruit. The idea is that each person learns how to make a contact microphone, and then you put the contact microphones onto the watermelons to turn them into percussion instruments, to facilitate a drum circle.

AG: So, the first part of the evening, you're actually making the device?

MA: Yeah.

AG: OK, so you arrive, and there's a bunch of wired watermelons.

MA: No, you would wire your watermelon.

AG: [Laughs] Right. So one of the things we've been trying to address is this impulse to popularize science and engineering through very hand-on and heuristic learning projects, like the recent Enormous Microscopic Evening you produced with CRITTER Salon at Hammer. What would you say are the contributing factors to this impulse to make things, to be very hands-on in terms of disseminating information? Instead of having a PowerPoint presentation, you make your own amplified watermelon.

MA: I think there's a variety of things. In general, my philosophical position is that cultural producers are symptoms rather than driving forces of what's happening. I would say what you're witnessing is recreational use of rhizomatic information networks (basically like people looking at shit on Wikipedia). And then there's a shift toward participatory culture, also driven by networks. And then these [online] tendencies manifest in the real world. I know that for me, as someone who spends a lot of time on the computer, I'm also profoundly interested in the bandwidth of what happens in real life. You realize that the highest bit stream of data [comes from] standing in front of somebody? You get body language, you get gesture, you get smells, you get pheromones, you get all of that. To try to summarize: a rise in participatory culture, an interest in how network culture has articulated the specificity of non-network cultural experiences, and the use of information as recreational entertainment. One other thing I think is important: there's very little in our daily lives which

is not constructed via a huge system. Whether it's where our food comes from, or how this paper plate I'm eating a bagel off of is manufactured, or soap or whatever. I've always thought our interest in technology is giving people a moral, conscious agency in relation to the constructed environment. That means how to make soap is just as important and mysterious as how to program a computer. So we've always tried to embrace technology in a very wide sense, meaning "that which constructs the world around us." I was very happy last year when we had a day where we had a Paleolithic series about technologies a caveman would need to know.

AG: [Laughs] This fits well within our timeline.

MA: We had a day where we were in the basement and we had 10 people making stone knives using obsidian and rocks, and then upstairs in the gallery there was an iPhone programming class. And in the apartment we were doing poetry readings for people. I like that kind of simultaneity.

AG: And the events were on multiple levels—almost archeological.

MA: Yeah, exactly.

AG: Or geological. Well, this is fantastic. I think this is confirmation of things we've been writing about. Thank you very much, Mark, for bending to our strange timeline.

THOMAS THWAITES
THE TOASTER PROJECT
2010

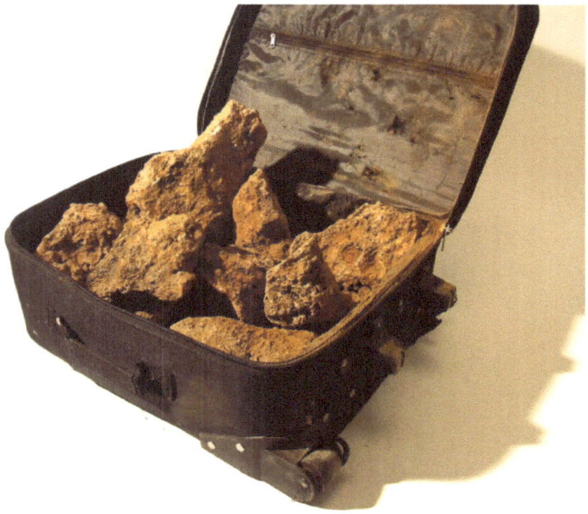

131.

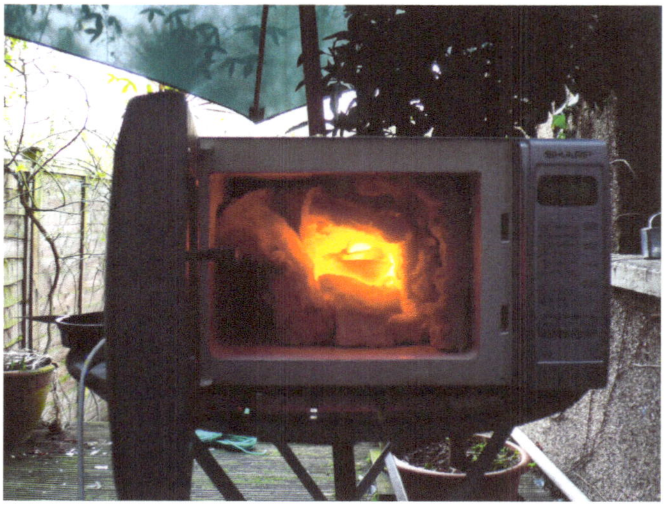

132.

131, 132. Images from Thomas Thwaites, ***The Toaster Project***, 2009.

The Toaster Project is an ambitious experiment in domestic reverse-engineering by the British designer Thomas Thwaites. Baffled by the fact that he could buy a toaster for a mere £3.99 at the supermarket chain Argos, Thwaites undertook an unusual experiment—to try and build a toaster from scratch. This, as it turned out, involved processes most people would never dream of attempting, like smelting iron ore, dug from an abandoned mine, in a household microwave.

A humorous reflection on issues of sustainability, industrialization, mass consumption, child labor, and DIY culture, Thwaites's finished product reflected the true cost of the modern global manufacturing complex behind an ostensibly cheap appliance: his toaster cost £1,187.54 and took him nine months to make.

Was this a ridiculous undertaking? Thwaites explains, "It depends on the scale at which you look. Looking close up, a desire (for toast) and the fulfilment of that desire is totally reasonable. Perhaps the majority of human activity can be reduced to a desire to make life more comfortable for ourselves and has thus far led to being able to buy a toaster for £3.99... but looking at toasters in relation to global industry, at a moment in time when the effects of our industry are no longer trivial... they seem unreasonable. I think our position is ambiguous—the scale of industry involved in making a toaster is ridiculous, but at the same time the chain of discoveries and small technological developments that occurred along the way make it entirely reasonable."[1]

1. http://www.thomasthwaites.com/thomas/toaster/page2.htm

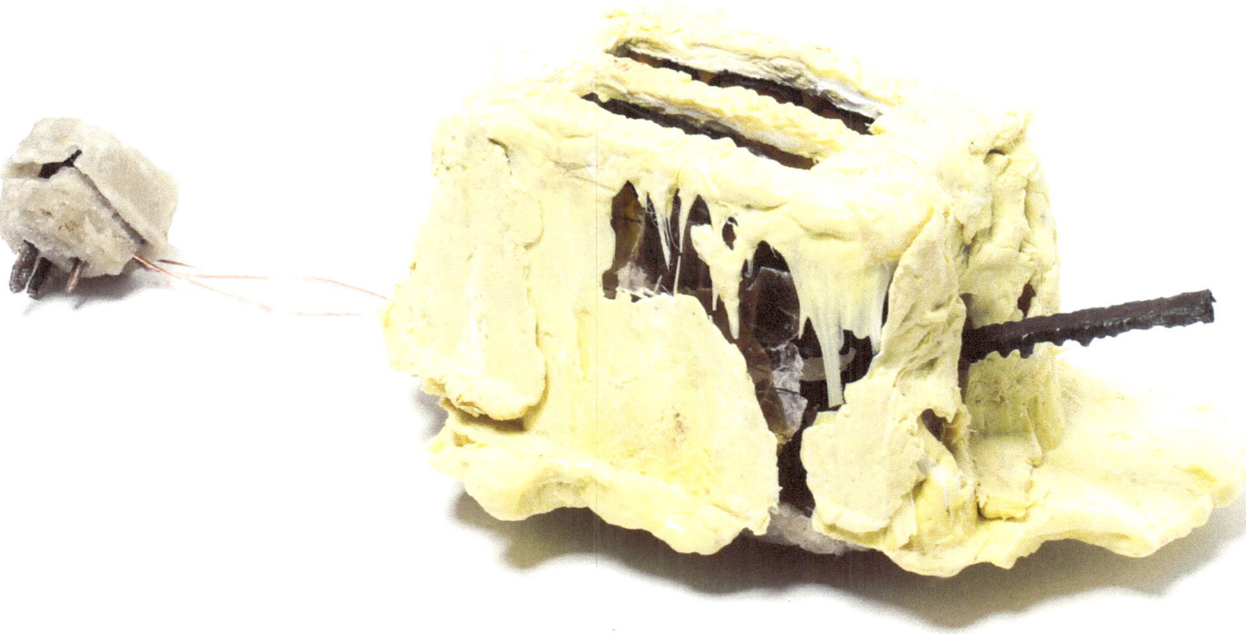

133.

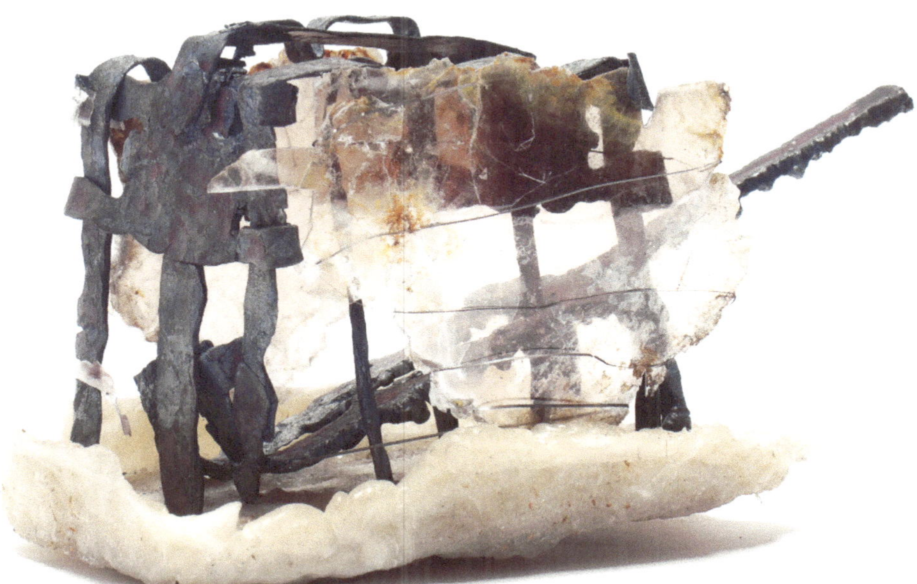

134.
133, 134. Images from Thomas Thwaites, *The Toaster Project*, 2009

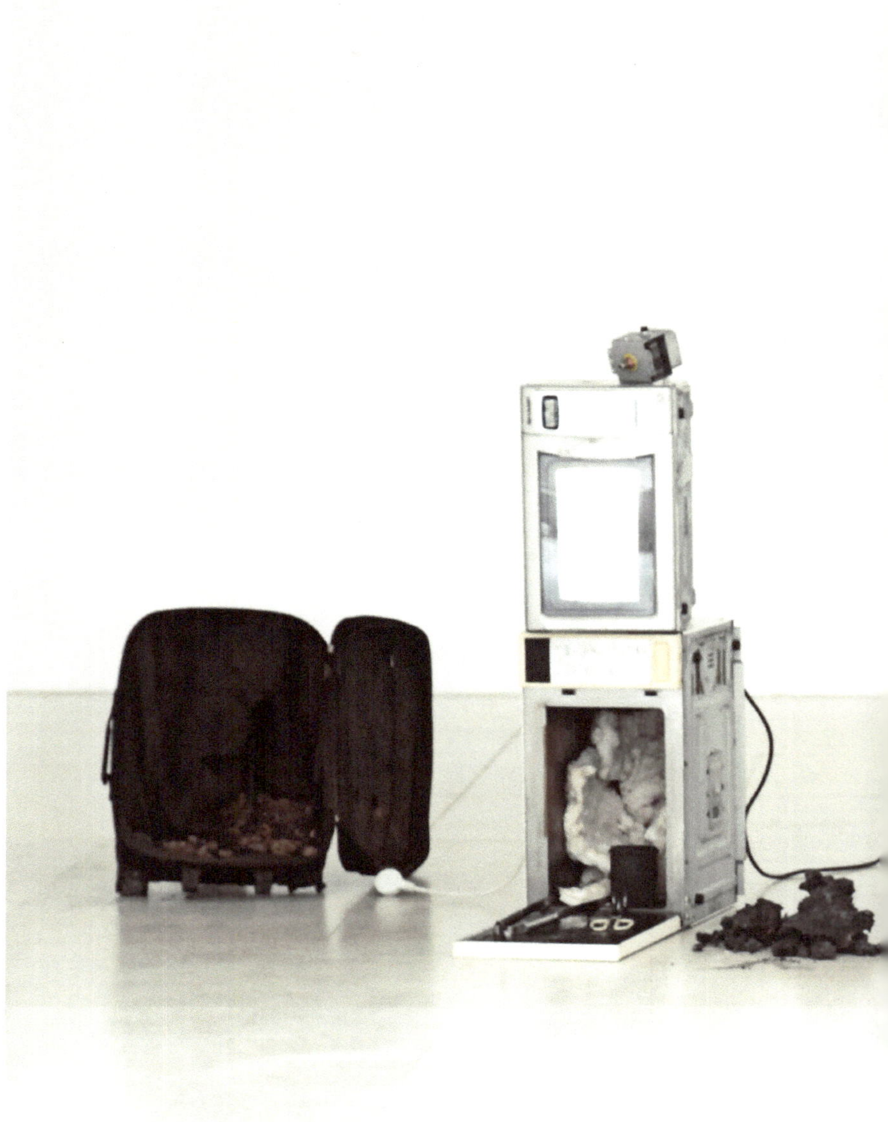

135. Thomas Thwaites, *The Toaster Project*, 2009

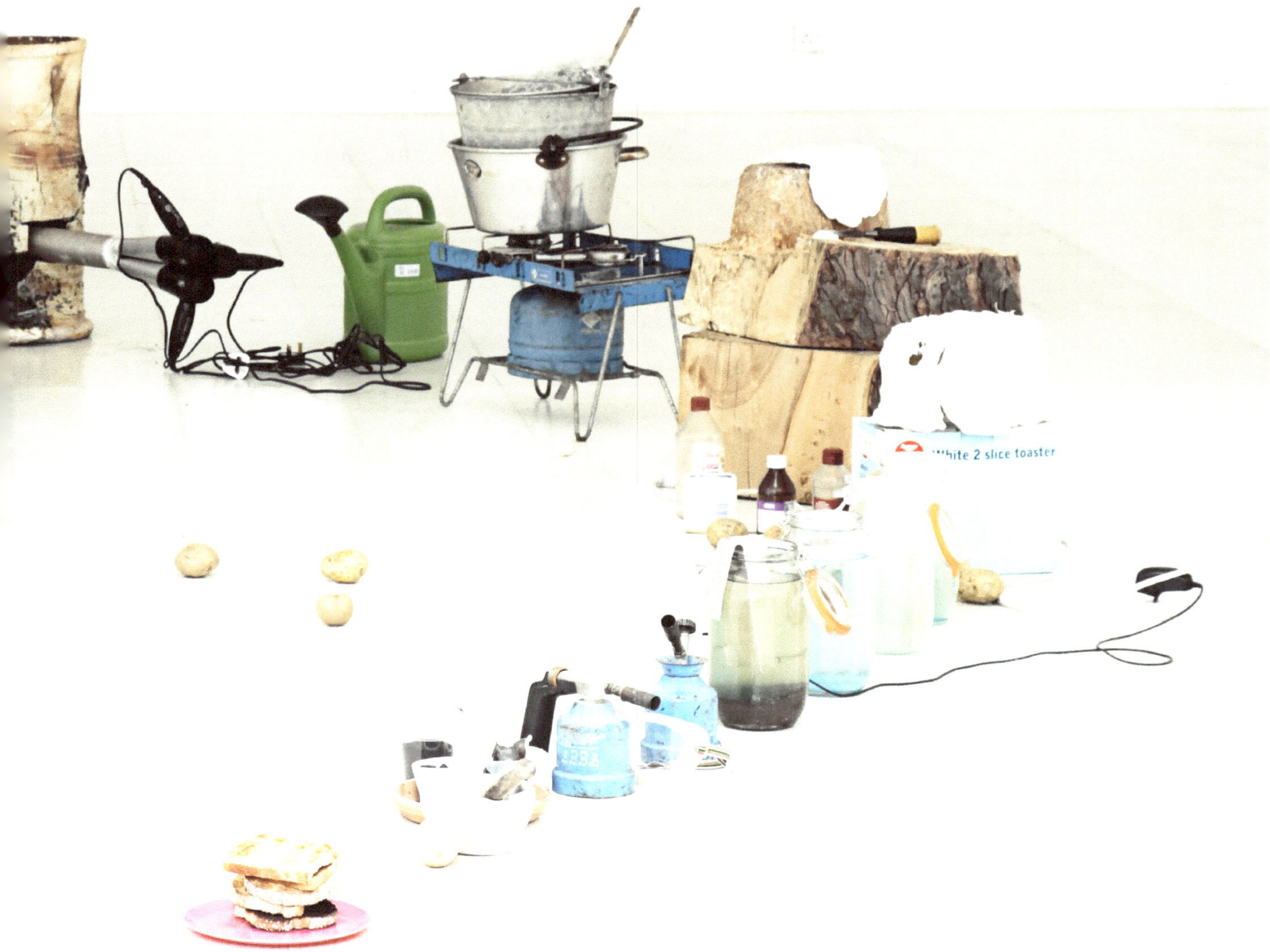
White 2 slice toaster

The Maker's Bill of Rights

Meaningful and specific parts lists shall be included.

Cases shall be easy to open.

Batteries shall be replaceable.

Special tools are allowed only for darn good reasons.

Profiting by selling expensive special tools is wrong, and not making special tools available is even worse.

Torx is OK; tamperproof is rarely OK.

Components, not entire subassemblies, shall be replaceable.

Consumables, like fuses and filters, shall be easy to access.

Circuit boards shall be commented.

Power from USB is good; power from proprietary power adapters is bad.

Standard connectors shall have pinouts defined.

If it snaps shut, it shall snap open.

Screws better than glues.

Docs and drivers shall have permalinks and shall reside for all perpetuity at archive.org

Ease of repair shall be a design ideal, not an afterthought.

Metric or standard, not both.

Shematics shall be included.

Drafted by Mister Jalopy, with assistance from Phillip Torrone and Simon Hill, from makezine.com

JONAH BRUCKER-COHEN AND KATHERINE MORIWAKI
SCRAPYARD CHALLENGE WORKSHOP
2010

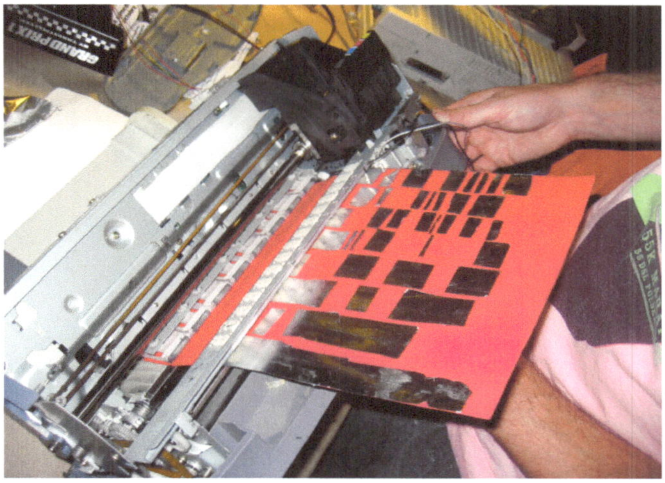

136. "Printer Drum Machine" from Scrapyard Challenge Workshop, OBORO, Montreal, Quebec, Canada, September 8, 2006

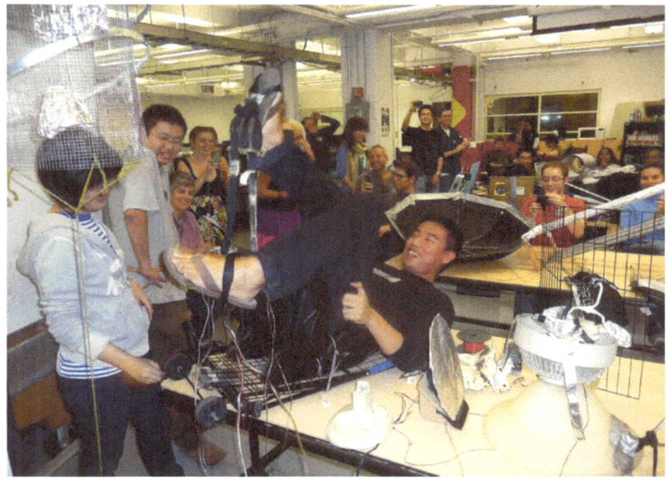

137. A Scrapyard Challenge Workshop at Parsons School of Design, 2010

Katherine Moriwaki is an assistant professor of media design in the School of Art, Media, and Technology at Parsons/The New School for Design in New York City. Jonah Brucker-Cohen is a researcher, artist, and writer with a Ph.D. in networking and telecommunications research from Trinity College Dublin. Together they run Scrapyard Challenge Workshops, intensive workshops in which participants transform discarded junk—maligned electronics, computer equipment past the point of its planned obsolescence, and so on—into innovative electronic projects. In short, hardware hacking. To date, these workshops have been held thirty-seven times in fourteen countries across five continents.

With themes ranging from the "MIDI Scrapyard Challenge" to the "DIY Wearable Challenge" (in the latter participants were defied to make wearable tech projects out of used clothing), Moriwaki and Brucker-Cohen's Scrapyard Challenge Workshops provide people with all levels of experience the opportunity to participate in interactive design. Scrapyard is about technology made accessible. As Moriwaki and Brucker-Cohen write, "The use of cast-off and cheap materials demystifies the idealized 'clean' aesthetic associated with technology...the explicit reference to the DIY and Hacker communities, as well as pop-cultural phenomenon, couch design practice in familiar layman terms."[1]

1. Jonah Brucker-Cohen and Katherine Moriwaki, "MIDI Scrapyard Challenge Workshops," presented at NIME 2007 (June 6-10, 2007).

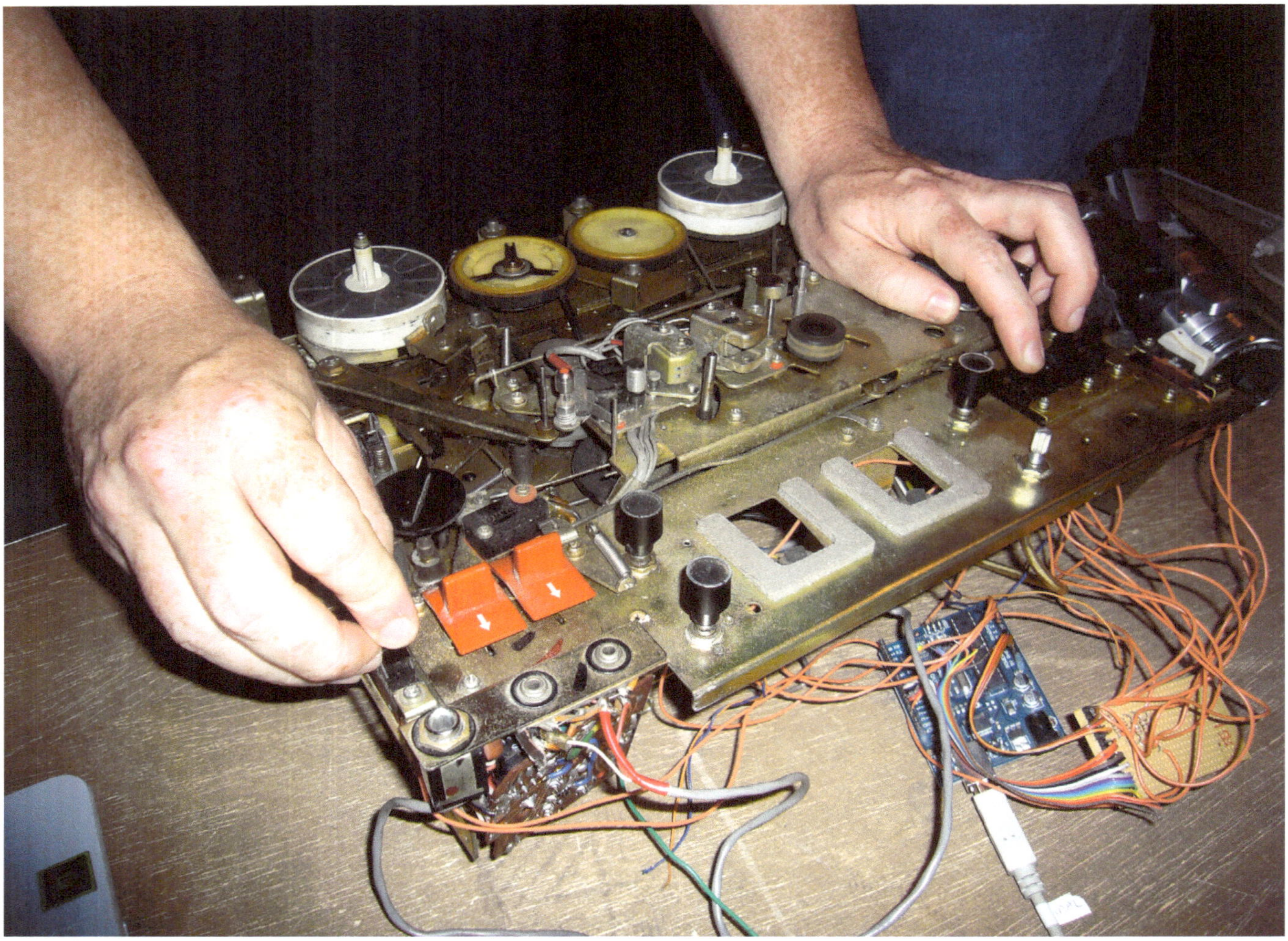

138. "Tapedeck Mixer" from Hybrid World Scrapyard Challenge, IMAL, Brussels, Belgium, 2008

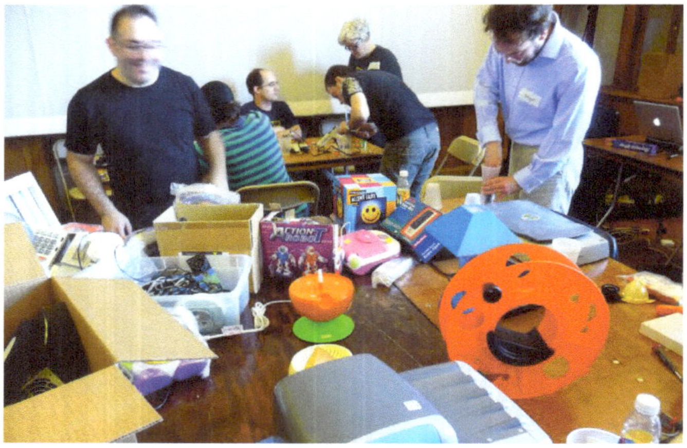

139. A Scrapyard Challenge at the STUDIO for Creative Inquiry, Carnegie Mellon University, November 2009

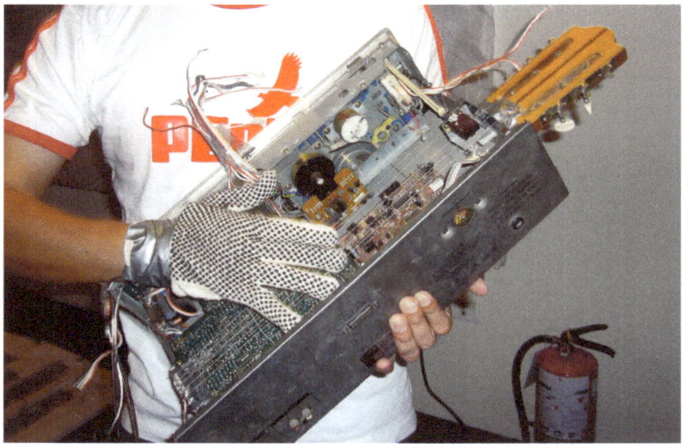

140. "Casemod Guitar" from Scrapyard Challenge Workshop, Fundacion Telefonica, _ima, Peru 11/5-9/07

Jonah Brucker-Cohen

As an artist, how do you see your role in a technological or scientific setting?

Since my work crosses the boundaries between art / design / and technology, there is definitely some crossover with scientific technological advancements. The Scrapyard Challenge workshops that I co-developed with Katherine Moriwaki, integrate hardware hacking and invention within a limited time frame to enable the creation of new interactive interfaces out of recycled materials and junk. This form of creation and invention involves rapid prototyping and experimentation which falls in line with methods carried out in the scientific community. Although our work might not discover new technologies, it reinforces the fact that through reusing existing technologies, there is a chance for a reimagining of interfaces and objects into newer, more complex forms than what was intended for their original use.

JON COHRS
URBAN PROSPECTING
2009

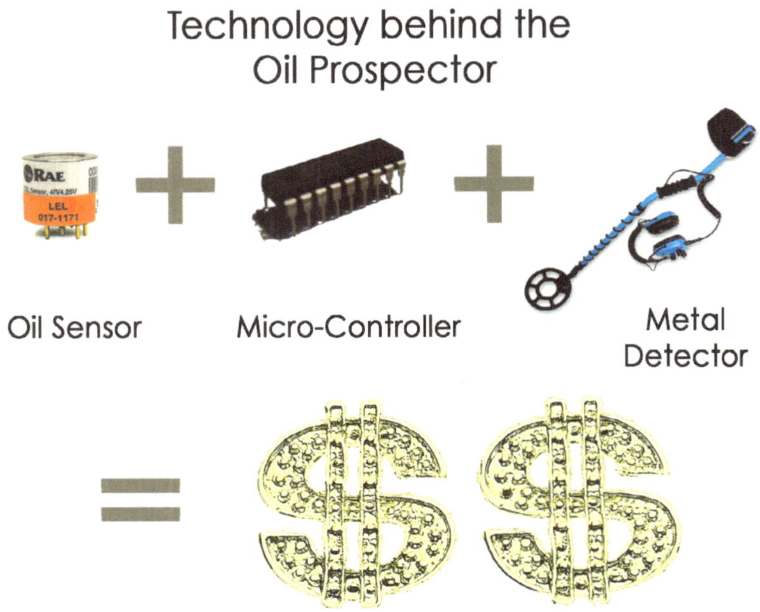

141. Technology behind Urban Prospector, Jon Cohrs, 2009

Inspired by a massive underground oil spill in Greenpoint[1], New York, that leaked seventeen million to thirty million gallons of oil and petroleum into the soil from crude oil processing facilities over a period of several decades, Jon Cohrs created a device, the *Urban Prospector*, that enables city dwellers to prospect for oil under the pavement and parks of their environment.

The *Urban Prospector* is a metal detector outfitted with a combustible gas sensor. The device can be built for under $100 following a set of instructions available online and using pieces anyone can buy secondhand or on eBay. By scanning the surface of one's neighborhood with this device, a prospector might spot pockets of oil, particularly in the vicinity of oil spills, abandoned gas stations, or industrial sites. Whole communities can use the tool to identify contaminated spots in their neighborhoods and map out areas of toxicity in a tangible grassroots way.

The *Urban Prospector* is DIY, satirical, and political. The project has been presented to the public as an opportunity to earn some cash on the black market. The *Urban Prospector* website declares: "Much like Gold, the value of Black Gold, or oil, has grown rapidly for the last 50 years. In many urban areas, industry has left behind resources that are written off as toxic spills. One can strike it rich simply by prospecting these industrial areas."[2]

1. For information on the Greenpoint oil spill, see http://en.wikipedia.org/wiki/Greenpoint_oil_spill
2. http://urbanprospecting.net/

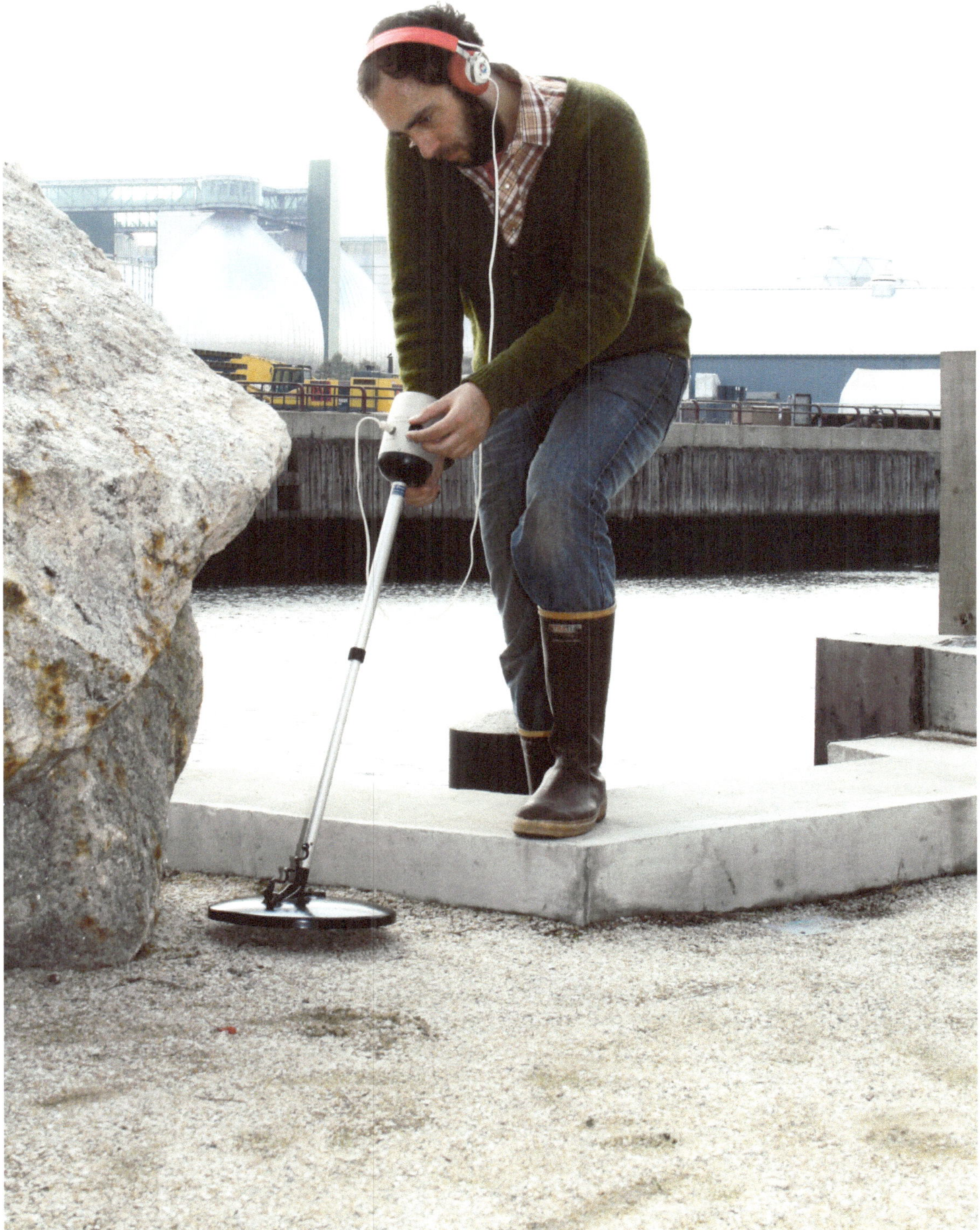

142. Urban Prospecting in Greenpoint, Brooklyn. Photo: Gaby Steiner, 2009

FREE ART AND TECHNOLOGY (F.A.T.), OPENFRAMEWORKS, THE GRAFFITI RESEARCH LAB, AND THE EBELING GROUP
THE EYEWRITER
2010

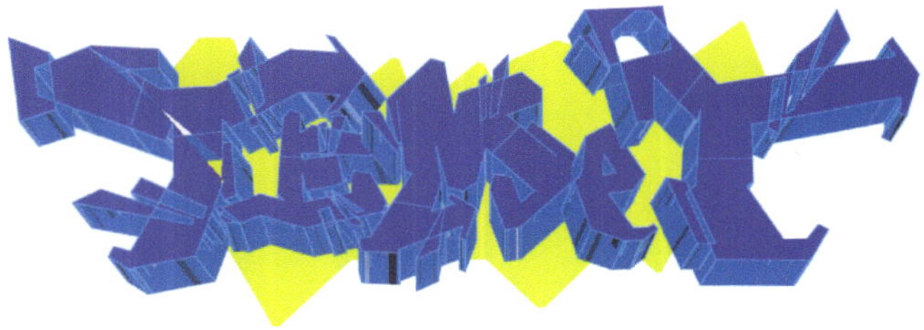

143. TEMPT1 tag, designed by the artist using his EyeWriter

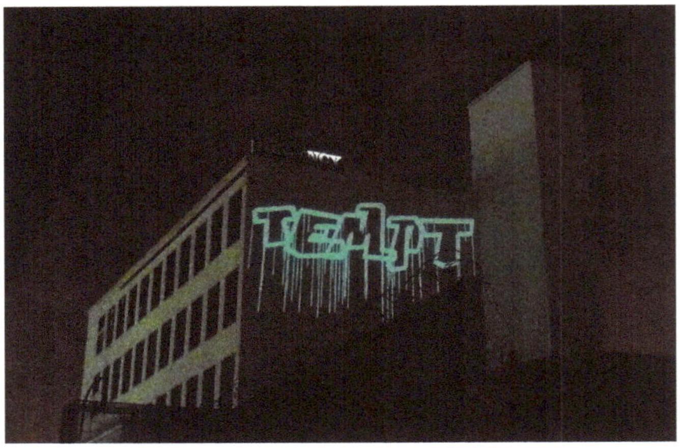

144. TEMPT1's EyeWriter tag projected live onto a building

In 2003, Los Angeles-based graffiti writer, publisher, and activist Tony Quan, aka TEMPT1, was diagnosed with late-stage Lou Gehrig's disease, or Amyotrophic Lateral Sclerosis (AML), which has left him almost completely physically paralyzed except for his eyes.

In response, members of Free Art & Technology (F.A.T.), OpenFrameworks, the Graffiti Research Lab, and The Ebeling Group worked together with TEMPT1 to develop an open source, low-cost eye-tracking device and software to allow the artist, as well as other ALS patients, to draw on a computer screen using only their eyes. Projecting the resulting images live onto the sides of buildings has enabled the paralyzed street artist back onto the street to engage in the larger dialogue of art, graffiti, and urbanism on a heretofore-unprecedented scale.

The long-term goal of the EyeWriter is to link together an international network of developers, hackers, and artists suffering from AML-induced paralysis to implement local materials and open source software in creating this device. In the open source spirit of the project, the team has published a tutorial on the website Instructables.com, enabling almost anyone with the time and materials to construct their own EyeWriter—the result is a proliferation of EyeWriter varieties by makers all over the world.

After seven years of immobility, TEMPT1 now uploads the graffiti designs he makes with this technology directly from his EyeWriter to a Flickr account online.

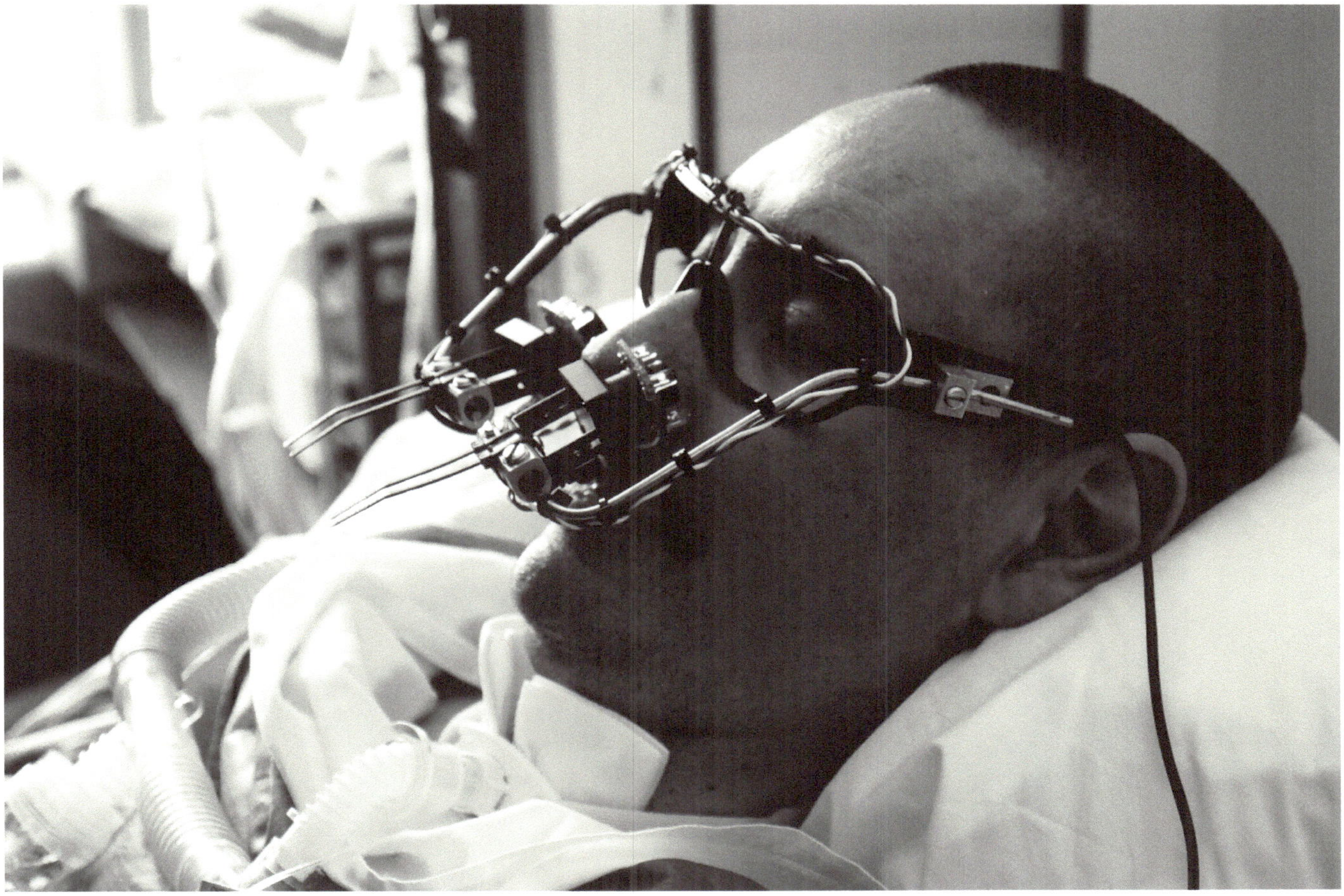

145. The EyeWriter, 2010

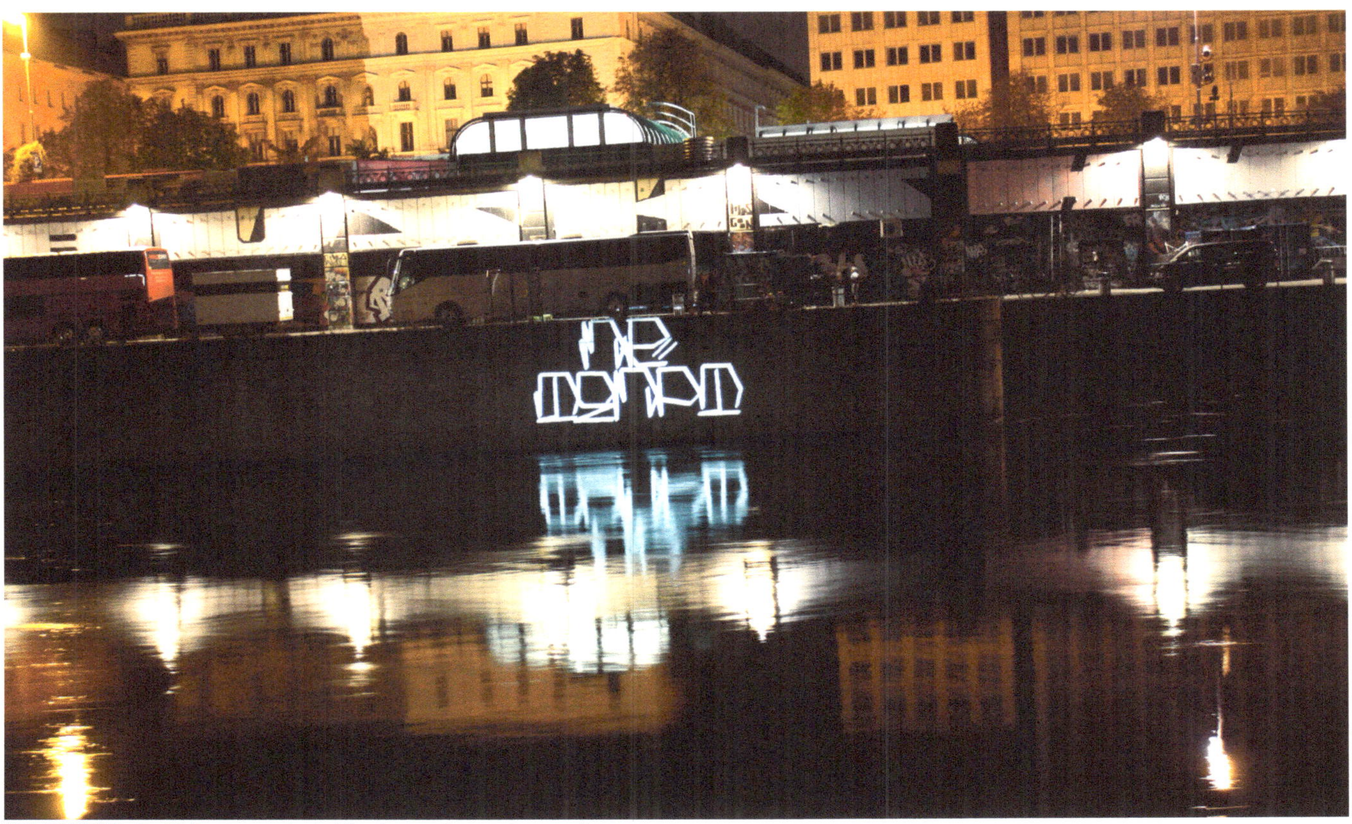

146. TEMPT1's EyeWriter tag projected live in an urban environment. More images available at http://eyewriter.org/

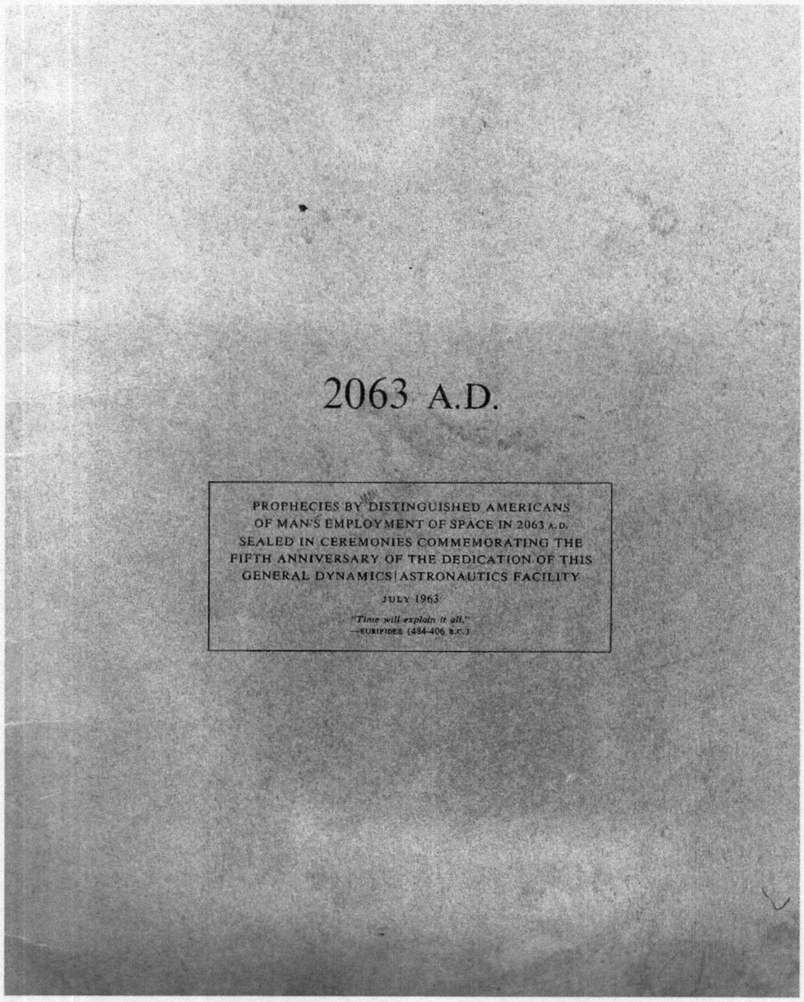
2063 A.D.

PROPHECIES BY DISTINGUISHED AMERICANS
OF MAN'S EMPLOYMENT OF SPACE IN 2063 A.D.
SEALED IN CEREMONIES COMMEMORATING THE
FIFTH ANNIVERSARY OF THE DEDICATION OF THIS
GENERAL DYNAMICS|ASTRONAUTICS FACILITY

JULY 1963

"Time will explain it all."
—EURIPIDES (484-406 B.C.)

147. General Dynamics 2063 AD Time Capsule

THE OVERVIEW EFFECT

ASPIRATIONAL, UTOPIAN, EXTRAPOLATIVE, SCIENCE FICTION, SYSTEMS, VISIONARY

When William Gibson used the term *cyberspace* in his 1984 novel *Neuromancer* to describe a "consensual hallucination experienced daily by billions of legitimate operators,"[1] the World Wide Web—and the complex web of relationships we experience within it—was not yet a fact of life. The word *cyberspace*, however, was immediately adopted to describe something that has developed over the years into an experience very close to what Gibson essentially described. This is the visionary approach: to make art that plays a role in disseminating and popularizing nascent and speculative concepts, anticipating cultural, social, and technological change.

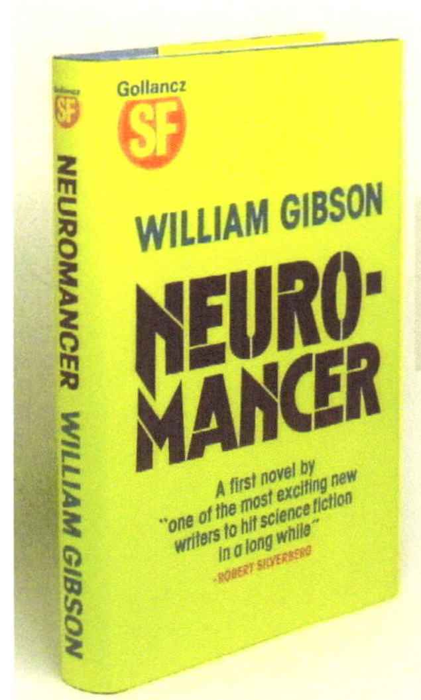

148. *Neuromancer*, by William Gibson. William Gibson foresaw both the existence of cyberspace and its use as a tool to ecstatically transcend the practical limits of reality

Isaac Asimov observed that science-fiction writers and readers didn't put a man on the Moon all by themselves, but they created a climate in which the goal of putting a man on the Moon became acceptable. Like science fiction, the work in this chapter is speculative and aspirational, and it views the current context as only a piece of a larger narrative, one that either reaches far into the future or operates within a philosophical framework of its own invention. One goal of this visionary approach is to create a climate in which the ambitions of the work can become accepted; the work is a living, breathing entity that informs the very world it critiques, inventing new myths, words, and realities just as we catch up to its old ones.

1. William Gibson, *Neuromancer* (New York: Ace Books, 1985).

This methodology has many points in common with design and architecture, by virtue of the fact that artists tend to translate philosophical and visionary ideas into livable environments and practical experiments. Historically, this visionary strand of art has been realized to great effect by hybrid architectural and environmental design practices like Superstudio, Archigram, and Ant Farm, which used a combination of hypothetical and practical projects to advance perception and set a precedent for a new understanding of the future. It also has roots in the "Design Science" ideas of R. Buckminster Fuller and the cosmogony of French Jesuit priest, geologist, and philosopher Pierre Teilhard de Chardin, who understood the "Noosphere" (the sphere of human thought) as moving toward an apogee of integration he called the "Omega Point." Teilhard de Chardin saw the goal of history as reaching an apex of consciousness, one that would culminate in the physical manifestation of thought. One could compare this aspiration more generally to the goals of artistic practice and specifically to this "visionary" methodology.

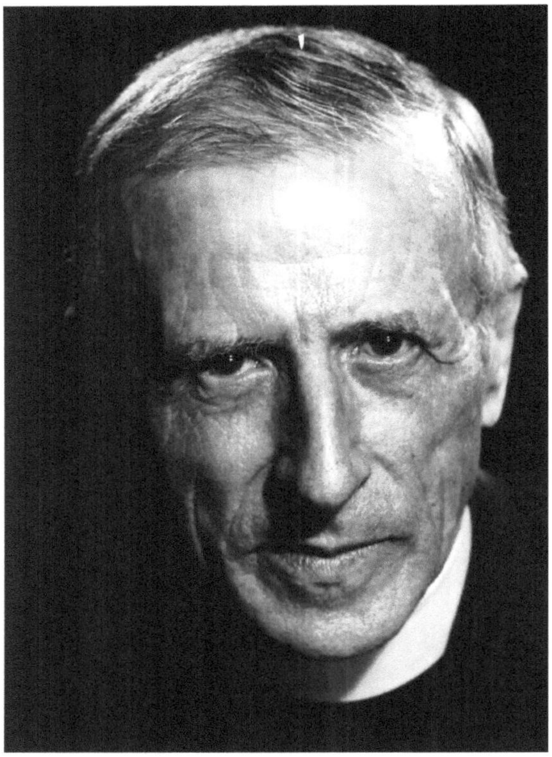

149. Pierre Teilhard de Chardin, trained as both a priest and a scientist, believed in the physicality of thought. He wrote, "Our duty, as men and women, is to proceed as if limits to our ability did not exist. We are collaborators in creation."

Lowry Burgess, an artist who has applied a version of this "visionary" methodology throughout his career, writes of a "metasphere" of "humanly created meaning... policies, technologies, and energies that surrounds the entire Earth and reaches into outer space,"[2] forcefully reshaping life on Earth. This is a divergent take on Teilhard de Chardin—one that the priest may never have thought of since he lived in a time that predated the emergence of the computer and global communications systems.

2. Lowry Burgess, quoted in *Living in Space: Cultural and Social Dynamics, Opportunities, and Challenges in Permanent Space Habitats*, edited by Sherry Bell and Langdon Morris (Aerospace Technology Working Group Books, 2009).

These are all essentially utopian ideas and movements—practices concerned with the creation of better or more sustainable environments for humankind—but "visionary" work needn't be entirely optimistic. Science fiction, for example, has a tendency toward dystopia, as it makes for a better story. Regardless, the shared objective lies in the radical discontinuity of a work from our present condition: a break or schism, a step backwards or away, a buffer of time or distance that affords us a fresh perspective on our reality, which—"forest for the trees"—we are not normally able to perceive. Important to the functioning of such a discontinuity is an element of familiarity: we see facets of ourselves projected forward or elsewhere, and it strikes us more powerfully than a pure fantasy would. Indeed, many cite this extrapolated, de-familiarized quality as a key conceptual difference between the broad categories of "science fiction" and "fantasy."

As the literary critic Robert Scholes wrote, "To live well in the present, to live decently and humanely, we must see into the future."[3] Of course, he could just as easily have written that "we must see into the past," as a study of history provides us with similar perspective—what is important is the disconnect, which can trigger an evolutionary change in thought. Take, for example, the "Overview Effect,"[4] that feeling of euphoria and profound connectedness experienced by astronauts looking back at the Earth from space. Although we all, technically, live in outer space, it takes this significant shift in point of view to realize the profundity of that truth. The philosophical and social implications in this instance are well documented—many argue that the first photograph of the Earth from space, "Earthrise," single-handedly triggered the environmentalist movements of the 1960s and 1970s. While to view from a distant vantage point the entire planet is perhaps the ultimate experience of radical discontinuity,

3. Robert Scholes, ***Structural Fabulation: An Essay on the Fiction of the Future*** (Notre Dame, Ind.: University of Notre Dame Press, 1975).
4. The "Overview Effect" is a phrase coined by writer Frank White, and explored in his 1987 book "The Overview Effect: Space Exploration and Human Evolution."

visionary artists can provide us with significant Earthbound fragments of this effect.

Other commonalities among the diverse practices detailed in this section are both their speculative nature and their propensity to deal in "big picture" ideas and holistic systems rather than isolated notions or events. Just as the science fiction writer envisions an entire world, the artist who enlists the use of this methodology models, visualizes, and explores a large slice of reality, if not the human condition itself.

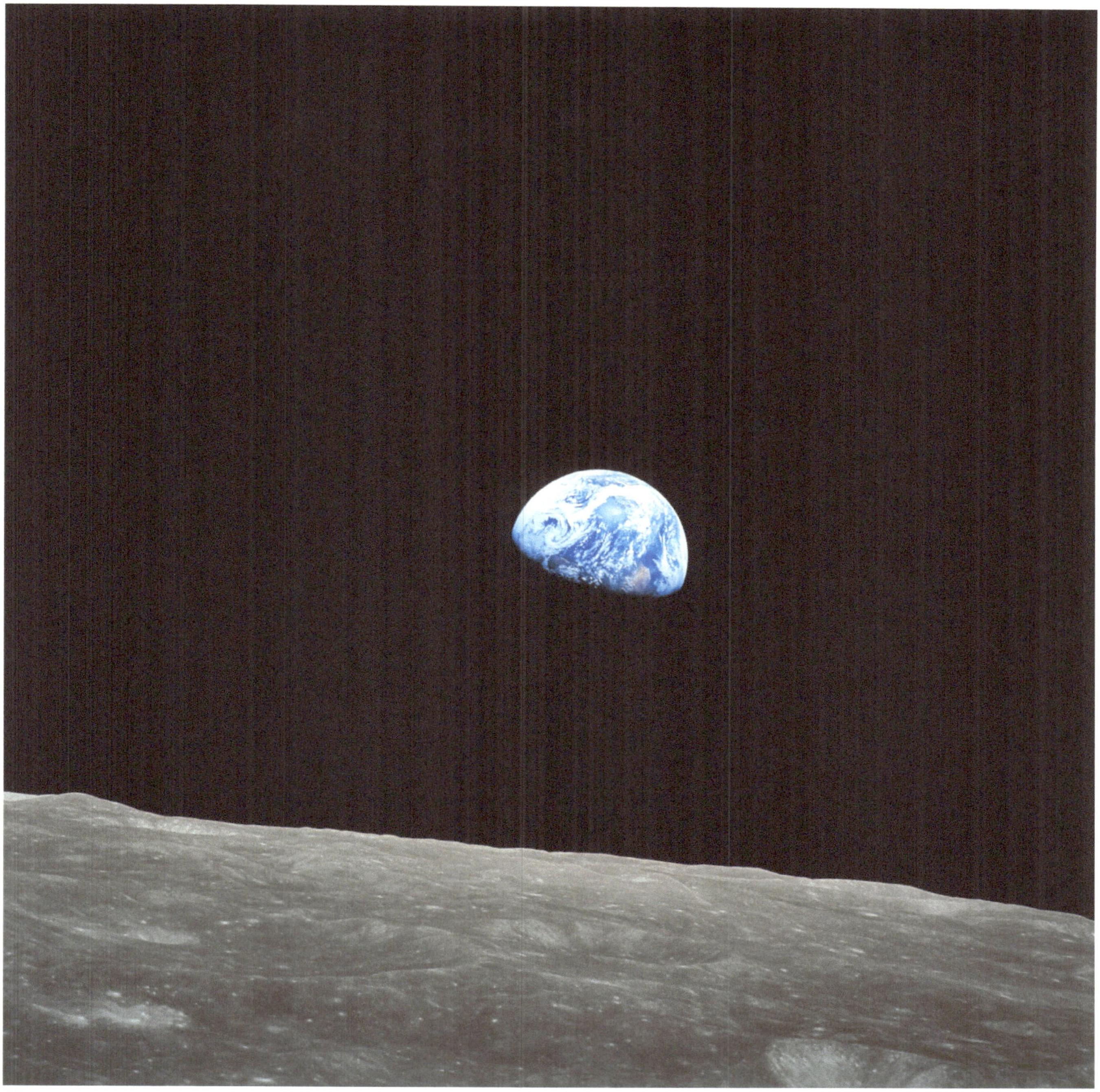

150. "Earthrise," snapped during the 1968 Apollo 8 Mission by astronaut William Anders, is often cited as one of the most influential photographs ever taken

TOMÁS SARACENO SUNNY DAY, AIR-PORT-CITY 2006

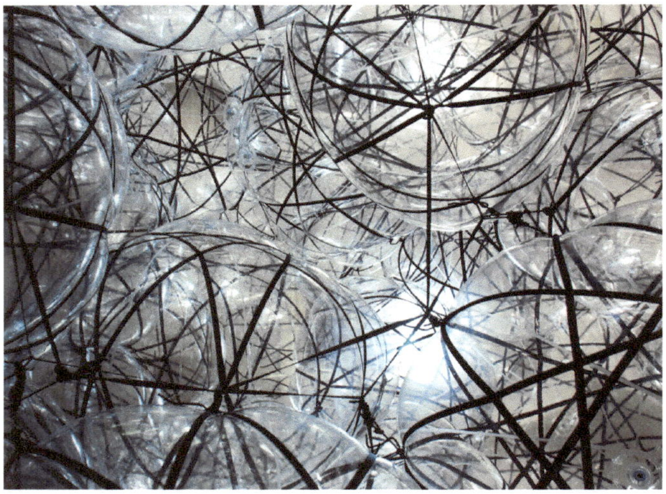

151. *32SW/Stay Green/Flying Garden/Air-Port-City*, 2007/2009, Tomás Saraceno. Courtesy the artist, Andersen's Contemporary, Tanya Bonakdar Gallery, and pinksummer contemporary art

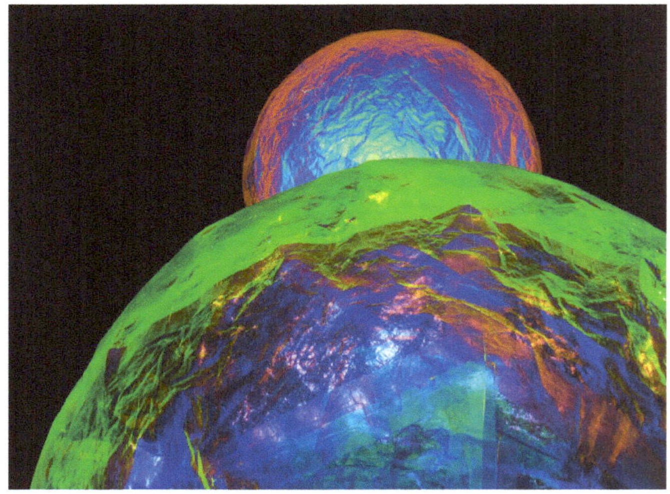

152. *Sunny Day, Airport City*, 2006, Tomás Saraceno. Courtesy the artist

The Argentinian artist Tomás Saraceno's 2006 *Air-Port-City* is a design for a mile-long geodesic "balloon city" that could navigate in international airspace and so provide an environment for people to live outside of national boundaries. Visualized with kinetic sculptures and inflatable spheres, the work proposes hospitable, solar-powered environments that float above the ground. In his practice, Saraceno borrows from scientific precedent and contemporary technological innovation to form ideas about new, sustainable communities and novel models for human connection. Aware of (and referencing) ecological considerations, as well as the interdependence of natural or biological systems, *Air-Port-City* is a self-contained system that exists within the natural world without necessarily interacting with it directly.

Saraceno's work has a utopian, collectivist quality that seems to stem from a lineage of aspirational building that includes the built and unbuilt structures of Archigram, Superstudio, Yona Friedman, Buckminster Fuller, or the collective known as Ant Farm. In a sense, *Air-Port-City* is a floating megastructure—an untethered piece of fantastical arcology. And yet, by virtue of the fact that it would float without restriction above the surface of the Earth, it challenges boundaries, nationhood, zoning, and all the structures of physical place that constrain the land-bound. It doesn't flaunt the rules, but rather exists outside of them in its own aspirational world; it exists in an alternate reality. In a radical disconnect, it provides its residents a view of the Earth from above the clouds—a mini overview effect.

About the piece, Saraceno has said, "My idea for an *Air-Port-City* is to create platforms or habitable cells made up of cities that float in the air. These change form and join together like clouds. This freedom of movement is borrowed from the orderly structure of airports, and it allows for the creation of the first "international city.... *Air-Port-City* is like a flying airport; you will be able to legally travel across the world.... This structure seeks to challenge today's political, social, cultural, and military restrictions in an attempt to reestablish new concepts of synergy."

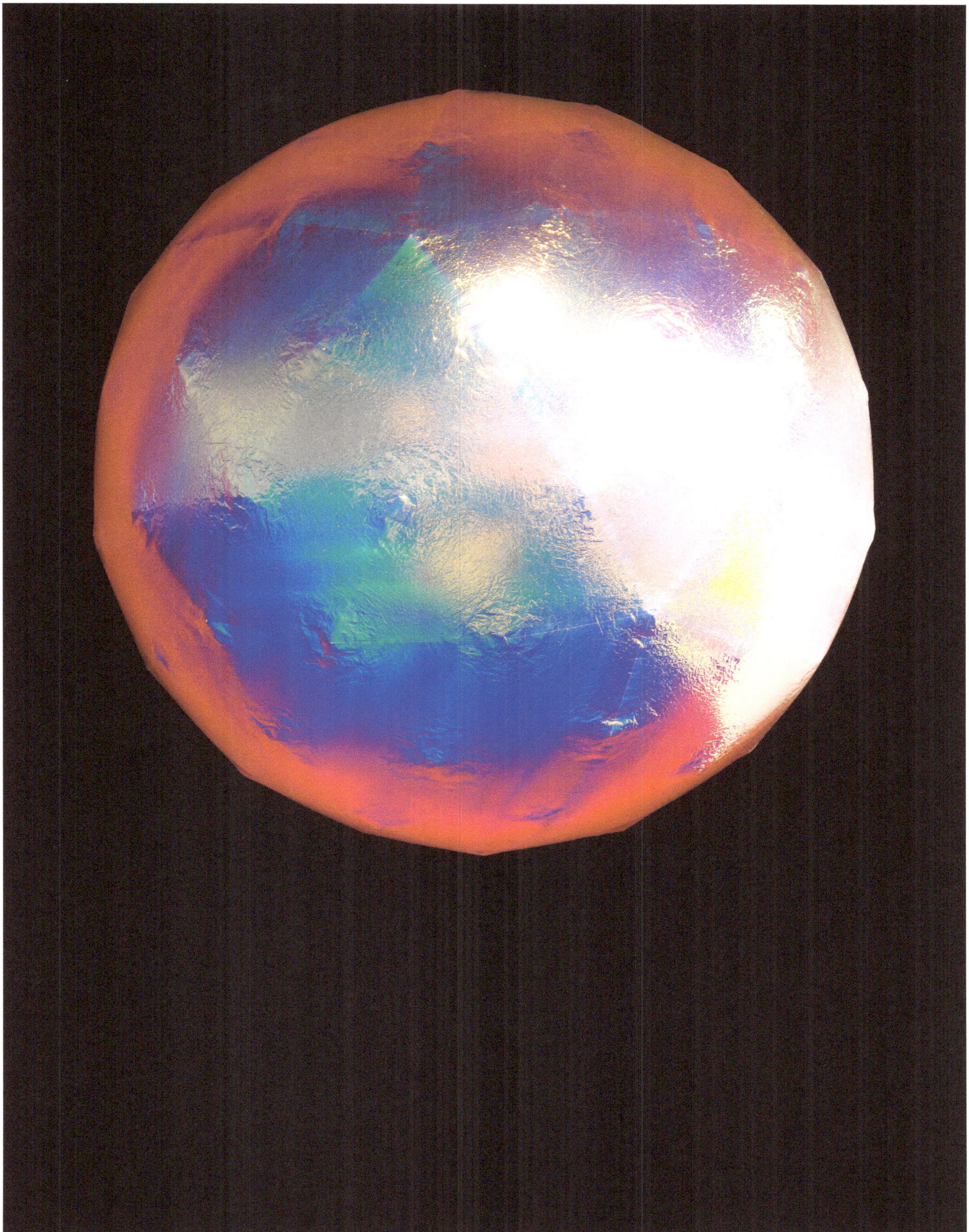

153. *Observatory/Air-Port-City*, 2008, Tomás Saraceno. Courtesy the artist and Tanya Bonakdar Gallery

Jeff Lieberman

As an artist, how do you see your role in a technological or scientific setting?

I see myself equally as an artist working in a scientific environment, and a scientist working in an artistic environment, having worked in both 'fields' relatively equally. Just as Leonardo da Vinci used pigments as his technology [and his knowledge of science to develop better pigments], I might use metal, hardware, and code — but only as tools in the service of expression of some form. I am not interested in technology on its own; I fell into that for several years; but technology has an inertia and momentum of its own, and is not often in the service of us as humans — it seems to separate us as often as bringing us together. I am only interested in the use of technologies to help us see our connectedness to the cosmos and to each other. I believe technology can be used to induce a sense of awe and wonder, an openness to new experience, in situations where people may be otherwise shut off. But it's most important what we do with that openness once it is triggered. If anything, I see my role as trying to help induce in people a sense of mystery in the universe around us, a mystery that pulls us toward each other, ultimately. As Neil Armstrong said, "Mystery creates wonder and wonder is the basis of man's desire to understand."

TOMÁS SARACENO
32SW STAY GREEN/FLYING GARDEN/AIR-PORT-CITY
2009

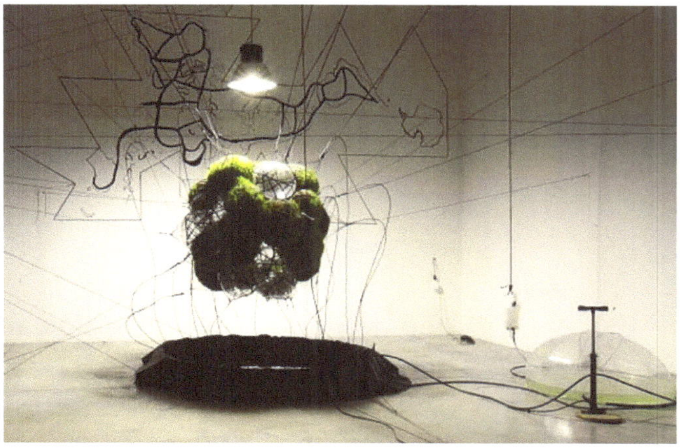

154. Tomás Saraceno, *32SW Stay Green/Flying Garden/Air-Port-City*, 2009

Solar panels placed throughout Tomás Saraceno's 2009 Walker Arts Center exhibition, "Lighter than Air," provided electricity to run an elaborate, self-sustainable greenhouse equipped with an irrigation system to water a cluster of inflatable spheres and green, living grass. This model for a "flying garden" dovetails with Saraceno's designs for an *Air-Port-City*; theoretically, the denizens of a floating balloon city could glean sustenance from this garden in the sky, alleviating the need to ever descend "back to Earth," both physically and metaphysically.

Practical designs like this are an important part of Saraceno's practice. Despite the idealistic tone of his work, Saraceno is serious about science. In researching for his spider web pieces, he worked with arachnologists, engineers, and astrophysicists to determine the tensile strength of webs; he has conducted research with NASA at the High-Altitude Platform, which uses aircraft as floating laboratories—our own proto-balloon city.

DUNNE & RABY
DESIGNS FOR AN OVERPOPULATED PLANET
2009

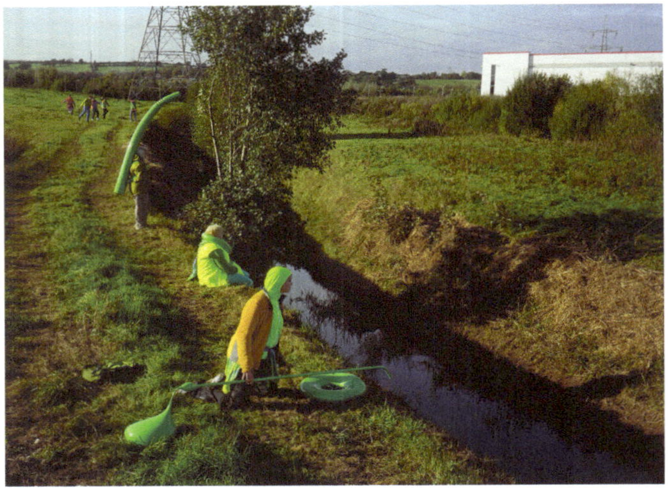

155.

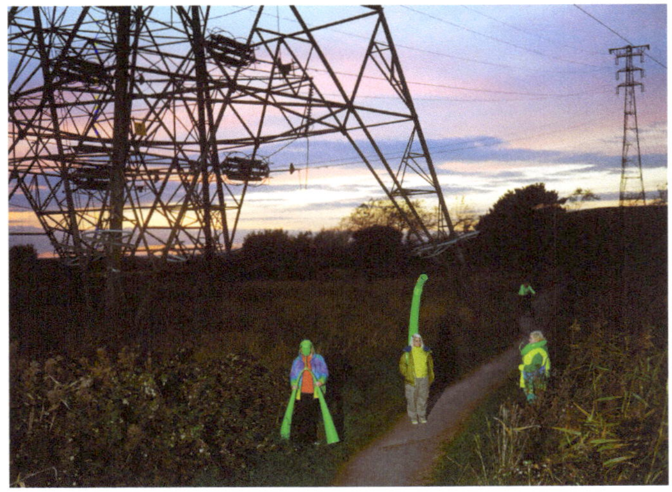

156.

155, 156. Dunne & Raby, *Designs for an Overpopulated Planet: Foragers*, 2009. Images courtesy Dunne & Raby. Photo: Jason Evans

Dunne & Raby form a London-based design studio that uses design as a medium for the catalysis of discussions about the implications—cultural, ethical, and social—of emergent technologies and scientific praxis. In *Designs for an Overpopulated Planet: Foragers*, they approach the idea of future food scarcity from a speculative biological perspective; instead of proposing practical solutions (as traditional designers might), Dunne & Raby propose a radical modification of the human organism. According to them, the project asks the question, "What if we could extract nutritional value from non-human foods using a combination of synthetic biology and new digestive devices inspired by digestive systems of other mammals, birds, fish, and insects?"

This is a speculative work, which includes ideas ranging from digestive clothing to designs for an augmented digestive apparatus that allows humans to consume tree branches, algae, or now inedible plants. Immediate practical application is not the goal. Rather, the project consists of models that can prompt reflection about the relationship between humans and their environment, both in the present and in the future.

It's also a narrative—a kind of science fiction told through design rather than language. *Designs for an Overpopulated Planet: Foragers* extrapolates an existing subset of society (food tinkerers, dumpster divers, and urban foragers) into a hypothetical future role that is not, in light of the rapid changes in biotechnology and the general direction of global development, entirely unrealistic. By placing a fictional group of DIY synthetic biologists and foragers at the center of the larger matrix of food scarcity, Dunne & Raby approach the massive problem of overpopulation from an unexpectedly speculative perspective—an extreme bottom-up approach. We can understand the "foragers" in this work as future versions of ourselves, people who have taken into their own hands responsibility for addressing the difficult situation that we, in our wastefulness, have bequeathed them. In displaying the radical methods by which we could theoretically extract nutrition from an urban environment, Dunne & Raby are designing solutions to consequences *that have not yet occurred.*

Dunne & Raby understand that our present use of natural resources is unsustainable, and this piece is as much a cautionary tale (essentially a dystopia: who wants to eat trees?) as it is a conjectural solution. Dunne & Raby refer to this kind of practice as "Design Fiction."

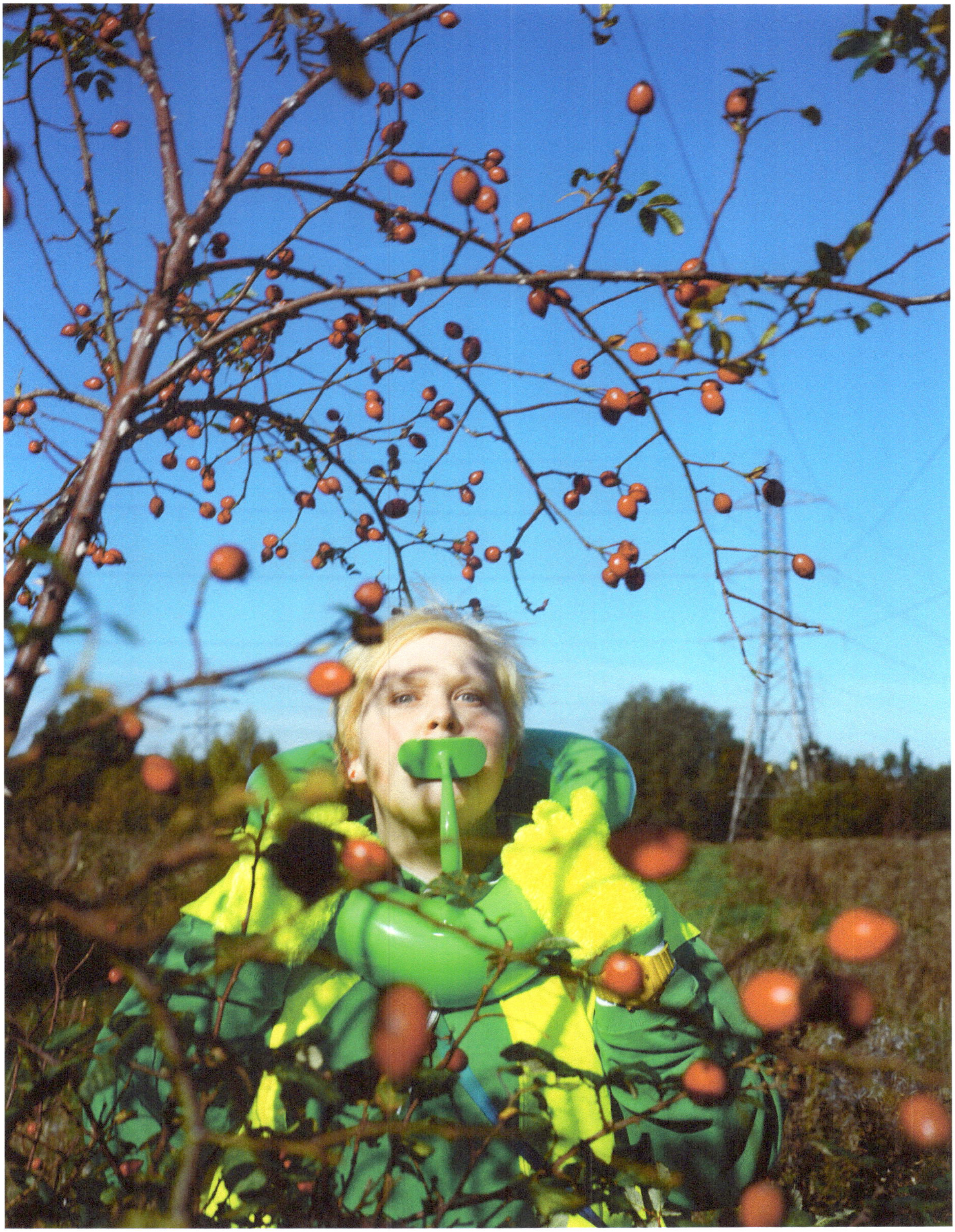

157. Dunne & Raby, *Designs for an Overpopulated Planet: Foragers*, 2009. Images courtesy Dunne & Raby. Photo: Jason Evans

October 18, 2010

Dear Mr. Kalil,

I cannot tell you how overjoyed I am that our Nation's leaders have finally opened an intrigued eye to the blossoming Maker movement. Your speech following the Maker Faire in New York was encouraging, exciting, and promising. It put a well deserved spotlight on the achievements of garage tinkerers and hackers around the country (and let's be honest, the world). That our leaders are paying attention to these atypical, underground activities and interested in turning them into mainstream, common American values is incredibly motivating to me as a maker.

There is, however, one facet of this movement that was overlooked in your speech, and as far as I can tell, is unfortunately overlooked everywhere STEM is championed. It is an undeniable aspect of humanity as valuable to Captain Picard as it was to Albert Einstein. It has been a driving force, technologically and economically, in the multi-billion dollar video game industry (and thus, the personal computer and home entertainment industries). It is introduced to Americans before Kindergarten, but somewhere along the path to high school, it is hopelessly abandoned as impractical and unproductive. But, it is also how we stop fragmenting ourselves into STEMs; it is how we come together to pick up STEAM for the renaissance (and yes, the ice cream was its idea).

Of course I am talking about art.

While art is a broad word that is dangerously all-encompasing (there is indeed an art to routing a circuit board, and a quite different art to designing a state machine), the art I am talking about here is fine art—that which Wikipedia defines as "developed primarily for aesthetics and/or concept rather than practical application." Fine art is no longer just painting on a canvas, drawing musical notes on a stave, or spinning clay into a pot. Fine art, in addition to everything it used to be, is electrical, dynamic, and algorithmic now, and to borrow from Oscar Wilde, as "quite useless" as it ever was. Take as an example, Syyn Labs' recent contribution to GLOW.

I wasn't at the Maker Faire in New York, but I have been to two in San Mateo, and many of the projects I saw there not practically useful, but were quite inspiring. Many of the useful projects I did see had one thing in common with the beloved MakerBots and DIYDrones: They were based on an Arduino, the open-source microcontroller and programming environment designed by artists for everyone.

Yes, the Arduino does fall under the blanket category of Technology, but it would be naive to think that its developers were trained only as technologists and engineers. Their training in art, sociology, and community is doubtlessly and inextricably linked to the platform's success across its diverse applications. Their desire to create something useful for artists is what drove them to simplify the user interface and lower the barrier to entry.

As I mentioned above, art also has a crucial role in the video game industry. Visual arts in video games are the reason why many of my friends own HDTVs. They are also the reason that companies like nVidia and ATi have had a thriving market in which to sell graphics cards and innovate parallel processing. By and large, people want the latest GeForce and Radeon cards for artistic reasons: they want their games to look good. It wasn't until very recently that using these massively parallel architectures for anything else was even reasonable.

I could go on about other examples of influences of fine arts on technology, like "Daisy" and the Altair 8800, but your time is valuable, and so is mine, so I'll cut to the point. This letter is to ask you to take a step back, have a look at the immense discrepancy between grant opportunities from the NSF and those from the NEA, and think about what we can do to pick up STEAM: Science, Technology, Engineering, Art and Math. Educating and encouraging our children to embrace artistic expression is just as important as teaching them calculus and the periodic table. Let's encourage our engineers to design new Most Useless Machines. Let's inspire our mathematicians to devise new mind-boggling N-dimensional fractal animations. Let's teach our artists to write programs and draw schematics so that they might create an electronic Mona Lisa. And let's show our children how fun and intertwined all of these fields are,so that they may form communities that flourish as they grow older and spread the joy to their children, and so on.

The train is headed in the right direction, we just need to invite everyone aboard.

Sincerely,
Kevin Nelson
Computer Engineer, Electronic Musician, Crasher

This open letter was written by Kevin Nelson in response to Thomas Kalil's (Deputy Director for Policy for the White House Office of Science and Technology) introductory remarks at the National Science Foundation sponsored conference, "Innovation, Education and Makers" and the promotion of STEM (Science, Technology, Engineering, and Mathematics Education Coalition).

SASCHA POHFLEPP
FOREVER FUTURE
2010

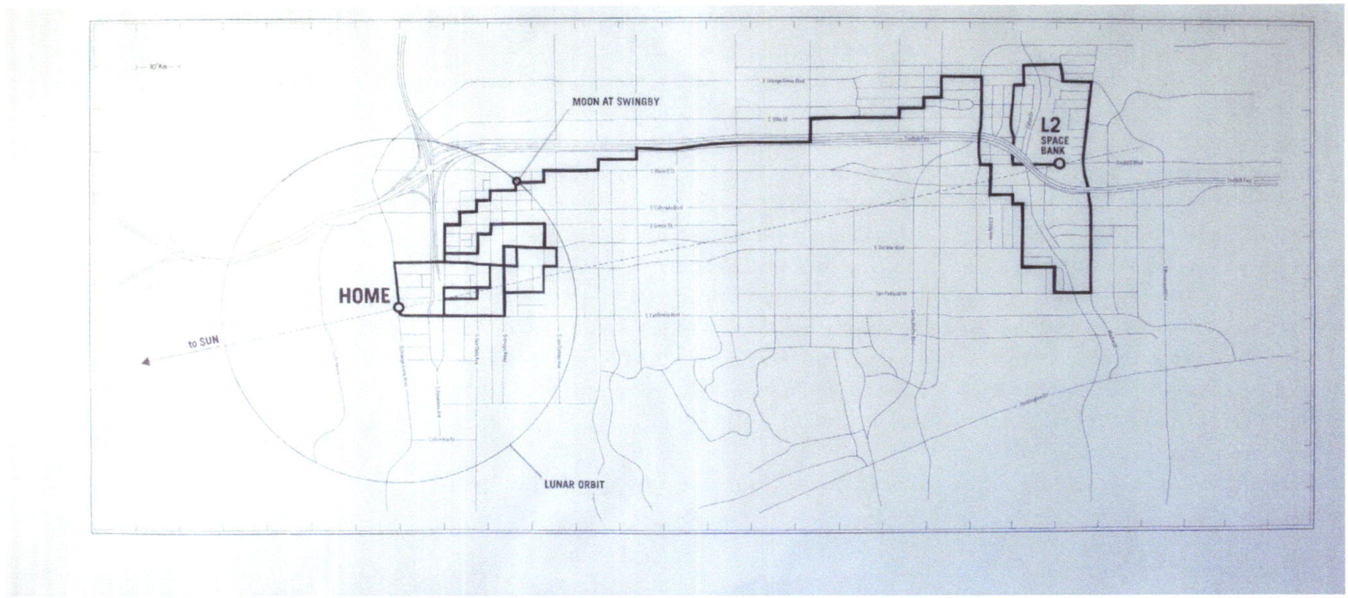

158. Sascha Pohflepp, *Forever Future*, "Walker's map for reaching escape velocity," 2010

Sascha Pohflepp's *Forever Future* is a video and installation work positing a fictional character named Robert Walker who suffers from nostalgia for "phantom futures," or those events predicted to happen within his lifetime that never actually materialized. Walker monitors the technological predictions made for each year, archiving those that don't come true and transferring them to 35mm slides. Every year, he makes a pilgrimage from his home to a storage unit in Los Angeles, where he deposits the materials into an object that vaguely resembles a space probe. Installed permanently in the storage unit and funded for perpetuity, this archive of unfulfilled dreams and predictions is designed to fly through *time* rather than *space*—at least until it is "recovered" by a future generation.

This work lays bare the social aspirations of technology: the desire to transcend the limitations of space, time, and materials. The present is complex and often frustrating, but the future is wide open—the site of dreams. Science and technology promise us space colonies, floating homes, increased leisure, and efficiency; we work toward a "better future" because we commonly understand it as an escape from the harshness of present-day reality. Pohflepp cites Jack Parsons, the rocket scientist and occultist, as an avatar of these ideas. Parsons sought both physical transcendence through his work with rockets at the Jet Propulsion Laboratory in Pasadena and spiritual transcendence through esoteric rituals, "sex magick," and a largely self-directed cosmology.

Forever Future encapsulates a holistic network of interconnected imaginary futures, which can coexist by virtue of their very *lack* of existence. Pohflepp's character, Robert Walker, is so disillusioned by the unavoidable truth that the future is perpetually becoming the present that he actually feels nostalgia for events that never occurred; his space probe, or ship, is like a paleo-futuristic scrapbook, where the fictional and real commingle in the mind of its archivist.

Pohflepp writes, "We do not know what happens when technological dreams don't come true, both on a cultural and on an individual basis. The assumption is that ideas, once they have been part of the public imagination, do not go away. They might go to another place we do not have an expression for, a cultural limbo from where they might be materialized at another point in time. This place might be shared with ideas from science fiction, a pool of possible futures that engineers and entrepreneurs are tapping into. There might, however, be futures that for various reasons may never materialize, which appear to be speeding away and thus stay at a certain distance from us."

Although Pohflepp never uses the phrase, the ship in *Forever Future* is a time capsule. The practical considerations of a real time capsule aren't necessary to this work, since *Forever Future* occurs within a narrative framework, as opposed to being an action taken by Pohflepp himself. This adds a layer of obfuscation to the work that makes it a kind of fiction, a whimsical approach to archivism. After all, the relationship Walker

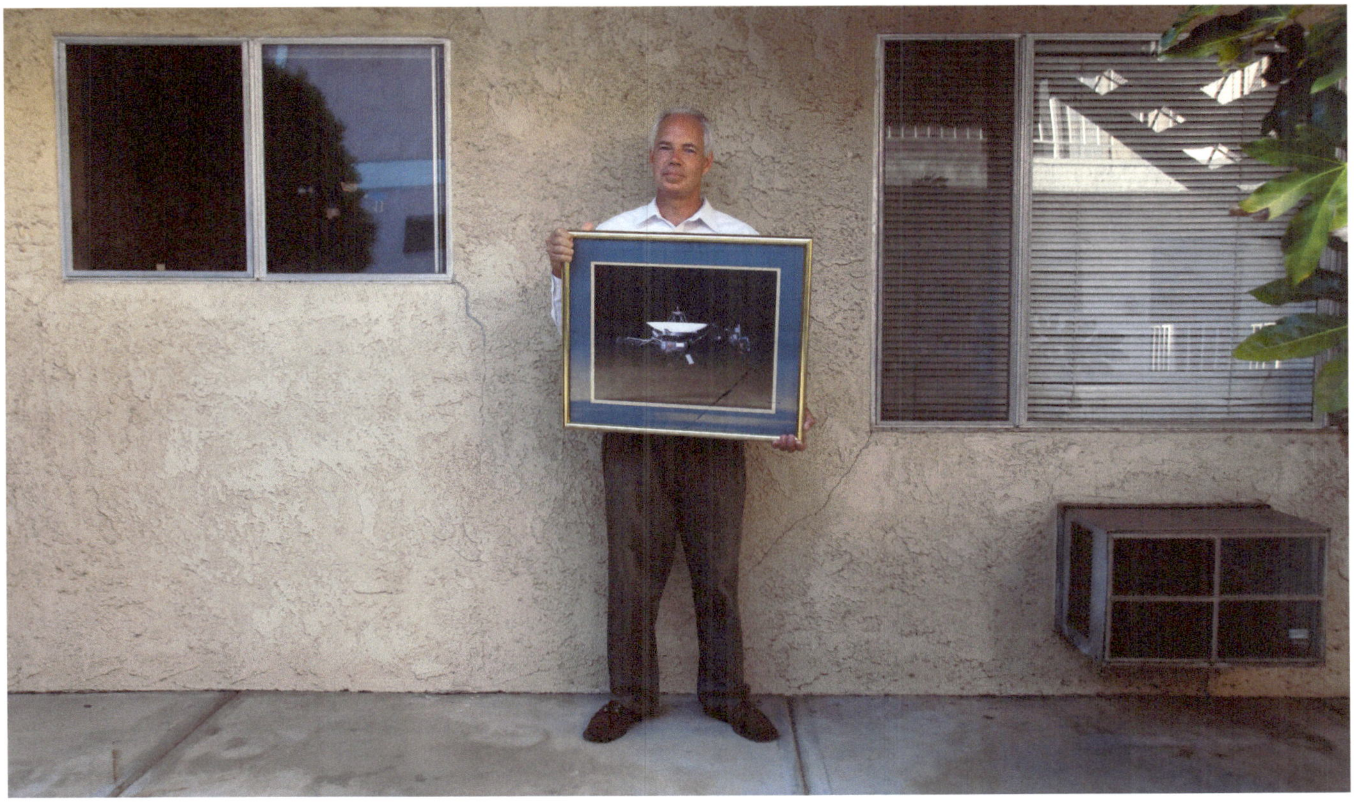

159. Sascha Pohflepp, *Forever Future*, 2010

has to the future mirrors the relationship that later discoverers of his work will necessarily have with the past. In other words, the "phantom futures" of Walker's archive, when viewed later, will have become phantom pasts.

The ship in Pohflepp's work is refrigerated, not unlike the time capsules built in the 1960s by Ant Farm, the radical architecture collective. Their capsules—sealed refrigerators or automobiles caulked in tar and buried underground—archived artifacts to be released into the not-so-distant future. However, the capsule they sent to the 6th Paris Biennal contained, among other things, souvenirs of the 1969 moon landing—items that Walker would have excised from his own archive.

Sascha Pohflepp

As an artist, how do you see your role in a technological or scientific setting?

The question about one's role is in fact keeping me busy a lot these days. Part of an attempt to answer it certainly lies in the question for why art and science have developed into such different beasts in the last couple centuries. We have certainly long passed the point where it was possible for a person to be a polymath, yet on the other hand many of the more recent developments in technology [that] promise levels of control over 'natural' systems have a purely knowledge-driven idea of science struggle. On a panel on 'picturing subjectivity', D. Graham Burnett pointed out that it is the "shared stances [of art & science] towards the world that are becoming part of how scientists and artists and engineers and designers in this more complicated space work within this idea of designed or made objects."

I do agree that there is huge potential and indeed demand for the attempt to bridge the divide between different ways of making sense of the world. Yet, I do also believe that there are certain qualities to each art and science which are difficult to translate and often end up in shallowness and masquerade with little contribution be it poetry or knowledge. It is a very difficult task, from the artist's perspective, to make something which both succeeds on the more intuitive cultural layers of what we value about art and the ones that are usually associated with science. Inversely, art that tries to be scientific rarely makes the jump from its fascination with the technological to the sublime and thus marginalizes itself as can arguably be seen within media art.

A possible escape from this conundrum could be for the artist to zoom out and try to understand how a new technology may be situated within the larger frameworks of history, society, economy, and so on – bridging divides between an atomized multitude of understandings while reaching beyond merely providing futures for technology. Such subjective synthesizing of the bigger picture into work may help us to grasp the many agendas at work within something that is indeed as world-shaping and contradictory as science and technology as it is extending itself into the world of living matter.

BRUCE STERLING
MIRRORSHADES
1989

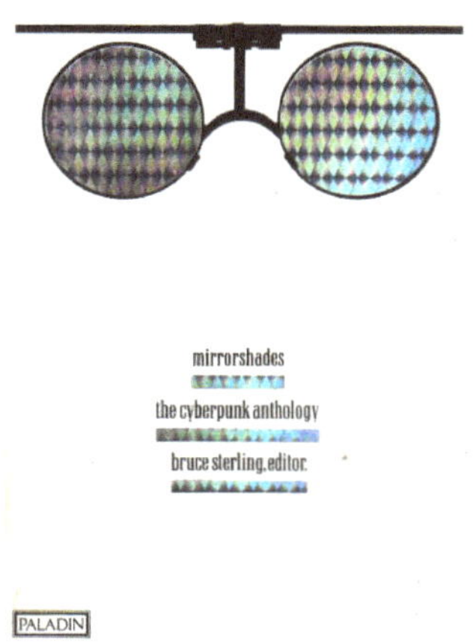

160. Cover image for *Mirrorshades: The Cyberpunk Anthology*, edited by Bruce Sterling (Paladin Books, 1989)

Bruce Sterling—sometimes known as "Chairman Bruce"—is a founding member and de facto spokesperson of the cyberpunk movement within the science fiction world. His contribution to the industry and craft of science fiction is significant: a multiple Hugo Award winner, he is one of the founding members of the Turkey City Writer's Workshop and is the author of dozens of novels and short story collections. He was the editor of the seminal cyberpunk anthology *Mirrorshades: A Cyberpunk Anthology*, which brought together writers like William Gibson, Paul Di Filippo, Greg Bear, and Sterling himself to define the then nascent genre.

Cyberpunk is science fiction that is keenly aware of how the exponential development of science and technology causes a foreshortening of the future. While older forms of science fiction (the so-called Golden Age) are set hundreds and sometimes thousands of years forward, allowing for a buffer of time between the present and the fictional future, cyberpunk is science fiction of the visceral *now*—it speaks to encroaching slums, the increasing integration of biology and technology, the degradation of flesh, political corruption, the corporatization of the world, and social disorder provoked by rapid changes in technological norms. The cyberpunks, who came to prominence in the mid-1980s, believed that science and technology were no longer disciplines of the academic or industrial establishment—the so-called ivory tower—nor were they separate from the everyday existence of lay people. Rather, technology and science were only getting *more* populist, ubiquitous, and personal. As Sterling wrote in the introduction to *Mirrorshades*, "for the cyberpunks... technology is visceral. It is not the bottled genie of remote Big Science boffins; it is pervasive, utterly intimate. Not outside us, but next to us. Under our skin; often, inside our minds."

Sterling's science doesn't belong to authorities or professionals; it's found embedded in the minds and bodies of the masses. Sterling writes of "brain-computer interfaces, artificial intelligence, neurochemistry—techniques radically redefining the nature of humanity, the nature of the self." This focus on implantation—technology close to the skin, cybernetics, hybridity—is a methodology we see again and again with artists who practice at the intersections of art, science, and technology.

After the heyday of cyberpunk, the movement eventually went the way of the Walkman and fell out of fashion, its decine in direct relation to its decreased relevance to current technology. However, its role as both predecessor and prophecy has not changed. Just as "punk" itself has become a lifestyle and ideology that goes beyond the initiating musical movement, Sterling has today begun to apply cyberpunk to more general writing about the state of the future, design, and environmentalism.

ATELIER VAN LIESHOUT
SLAVECITY
2005-2007

Atelier van Lieshout is a Dutch multidisciplinary company that operates in various fields, including (but not limited to) contemporary art, design, and architecture. Like many of the artists featured in this section, their work is speculative in nature, but *unlike* many of them, these speculations are not of a brighter tomorrow. Quite the opposite. Perhaps their most ambitious project, *SlaveCity,* is essentially a comprehensive design for dystopia.

The plan for *SlaveCity*, which includes designs for almost everything a functional city needs—from modular brothels and sanitation facilities to a "Minimal Steel Female University"—is exceedingly rational, efficient, and hence profitable. Atelier van Lieshout claims it would net a seven billion Euro profit annually. This comes, however, at the cost of human freedom: the 200,000 inhabitants of *SlaveCity* are referred to exclusively as "participants," spend their lives working fourteen-hour days at both manual and office labor, and are recycled into fuel (or, even more abhorrently, food) upon their death.

Of these participants' admission to *SlaveCity*, Atelier van Lieshout write: "Before entering *SlaveCity* and becoming an inhabitant of this city, you have to pass the Welcoming Center. In this large building, the participants are selected for their suitability to come and work in *SlaveCity*. Old, crippled, sick, and bad tasting people will be recycled in the biogas digester. Healthy, not so clever people will be recycled in the meat processing factory. Young and very healthy people will be able to take part in the organ transplant program. Healthy, clever people will go to work in the CallCenter."[1]

SlaveCity, with its similarities to the 1973 American film *Soylent Green*, in which food scarcity has led to institutionalized cannibalism, is a piece of sinister speculative design fiction. There is an eco-Orwellian rationalism to it; after all, *SlaveCity* recycles everything, including human waste and corpses. While designers like Dunne & Raby produce "Design Fictions" that present us with the consequences of present-day wastefulness, Atelier van Lieshout takes the opposite tack, drawing out latent fascist undertones from the green movement. We cannot help but compare *SlaveCity* to our environment. What path, which mistakes, which decisions might lead us to such a place? Have those choices already been made? Such a dystopia is as effective as work that is ambitious in a more "positive" direction: it shows us what our world might become in a purely logical future.

1. http://www.ateliervanlieshout.com/

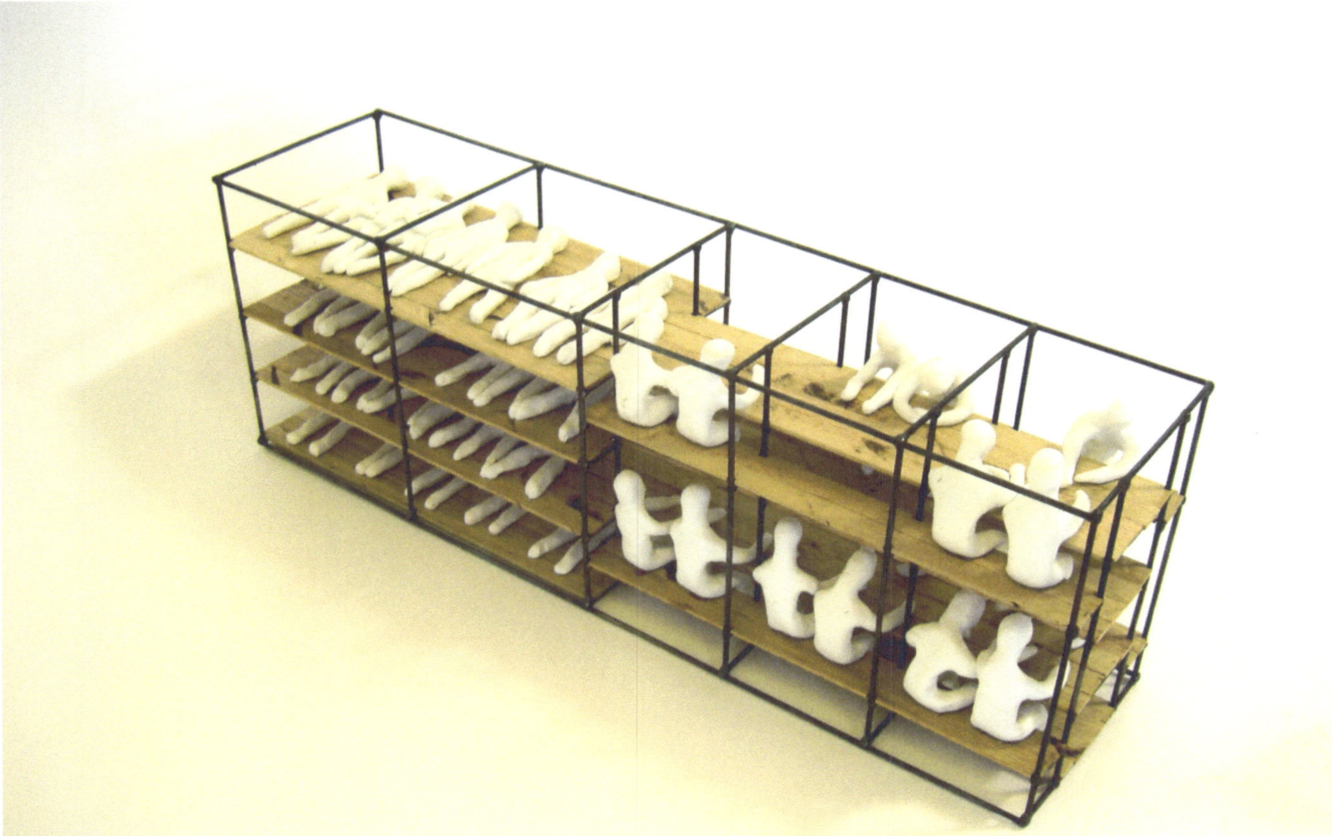

161. Atelier Van Lieshout, *SlaveCity – Model Sleepwork Units with Puppets*, 2005, photo & copyright: Atelier Van Lieshout

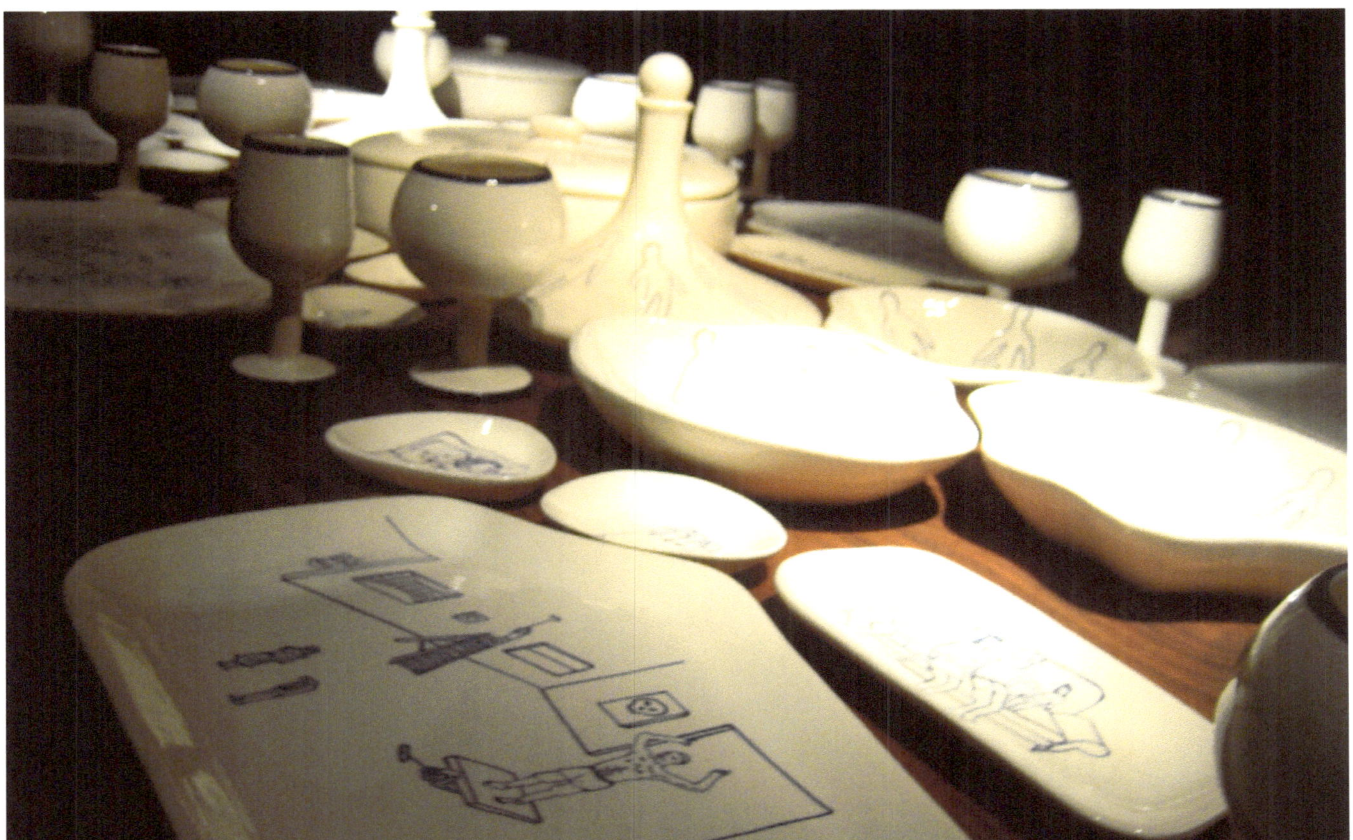

162. Atelier Van Lieshout, *SlaveCity - Board Room*, 2007, photo & copyright: Atelier Van Lieshout

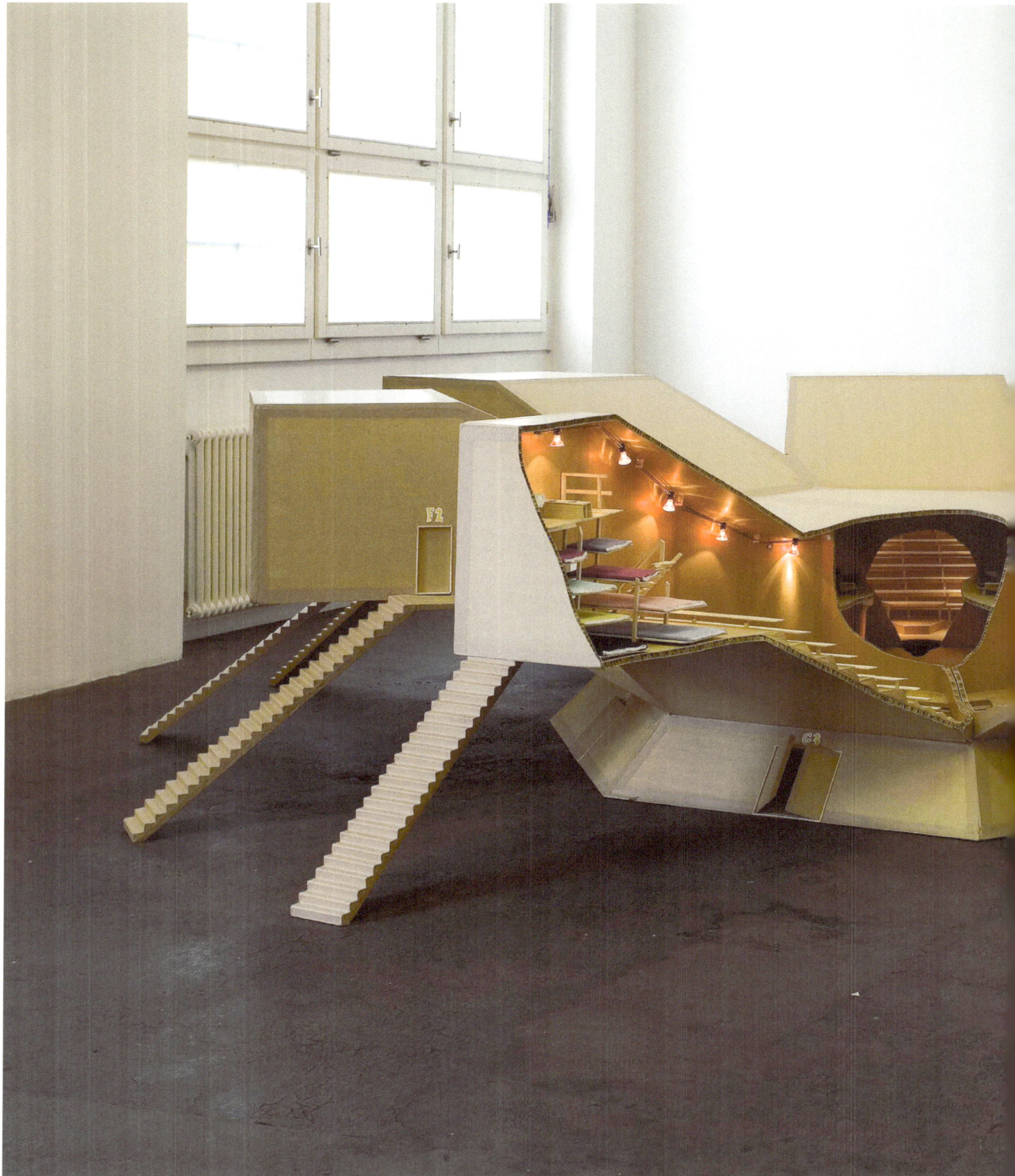

163. Atelier Van Lieshout, *SlaveCity – Male Slave University*, 2007, photo & copyright: Atelier Van Lieshout

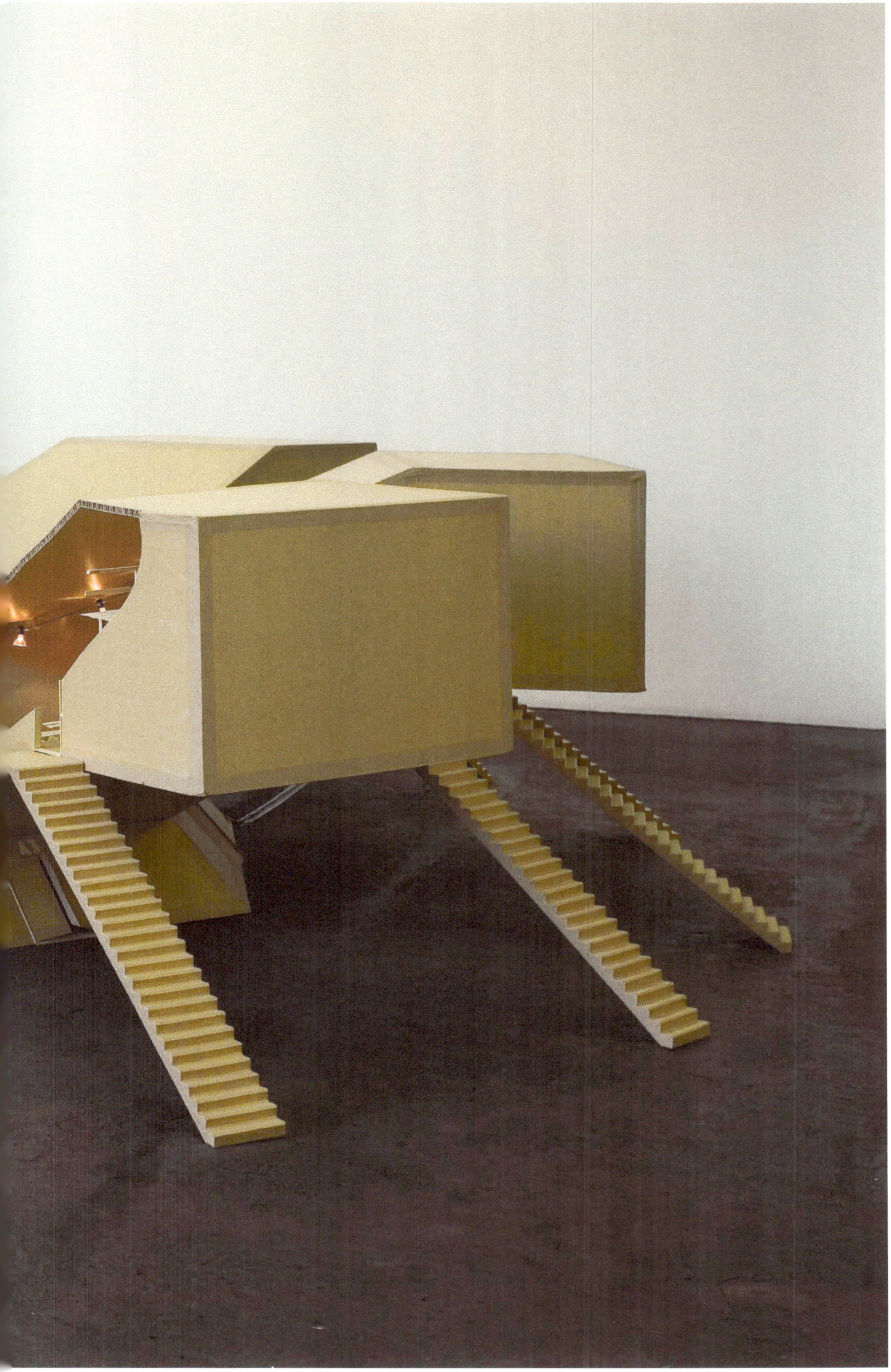

etoy.CORPORATION MISSION ETERNITY 2007-ONGOING

MISSION ETERNITY is a project currently being undertaken by the European digital art group etoy. etoy, whose primary slogan is "leaving reality behind," is a kind of art corporation that is traded and managed by its shareholders, who have a long history of performing complex and often litigious public performances. etoy is famous for its "toy war" with the online toy merchant Etoys.com, sometimes called "the most expensive performance in art history." Etoys.com, the toy company, fearing the confusion of its customers, attempted to shut down Etoy's website—the result was an electronic struggle and "tactical media event" waged by etoy and its participants against the company. etoys.com eventually dropped the legal dispute.

Their most recent undertaking, *MISSION ETERNITY*, is a "metaphysical adventure" exploring the relationships among death, identity, memory, and digital space. As much about loss as it is about the conservation of information, *MISSION ETERNITY* attempts to digitally capture people in "M∞ ARCANUM CAPSULES," which are interactive portraits—digital sarcophagi containing fragments of the life and soul of a person, stored when they are facing death. It is etoy's postulation that a person can be kept alive for eternity as "infinite data particles" that are "forever circulat[ing] the global info sphere." Much as we now live our lives increasingly online, we can continue to live after death online.

There are many components to the project. It claims to enlist the participation of hackers, scientists, and artists in an international effort that is technical, legal, spiritual, and economic. The project also calls on the public to serve as "M∞ ANGELS" by donating portions of their digital storage capacity to these "capsules," essentially maintaining the physical presence of the dead as data on a hard drive, mobile device, or network cloud. etoy explains that "the dead continue to exist as biomass and traces in the global memory: in governmental data bases, in family archives, in professional records, and in emotional data stored as electrical impulses in the bio-memory of our social network."

etoy's scope is large, and their vision, long. They have invested decades of their time into the development of *MISSION ETERNITY* in order to construct "a bridge between life and beyond: a trip into the past, presence, and far future of human culture."

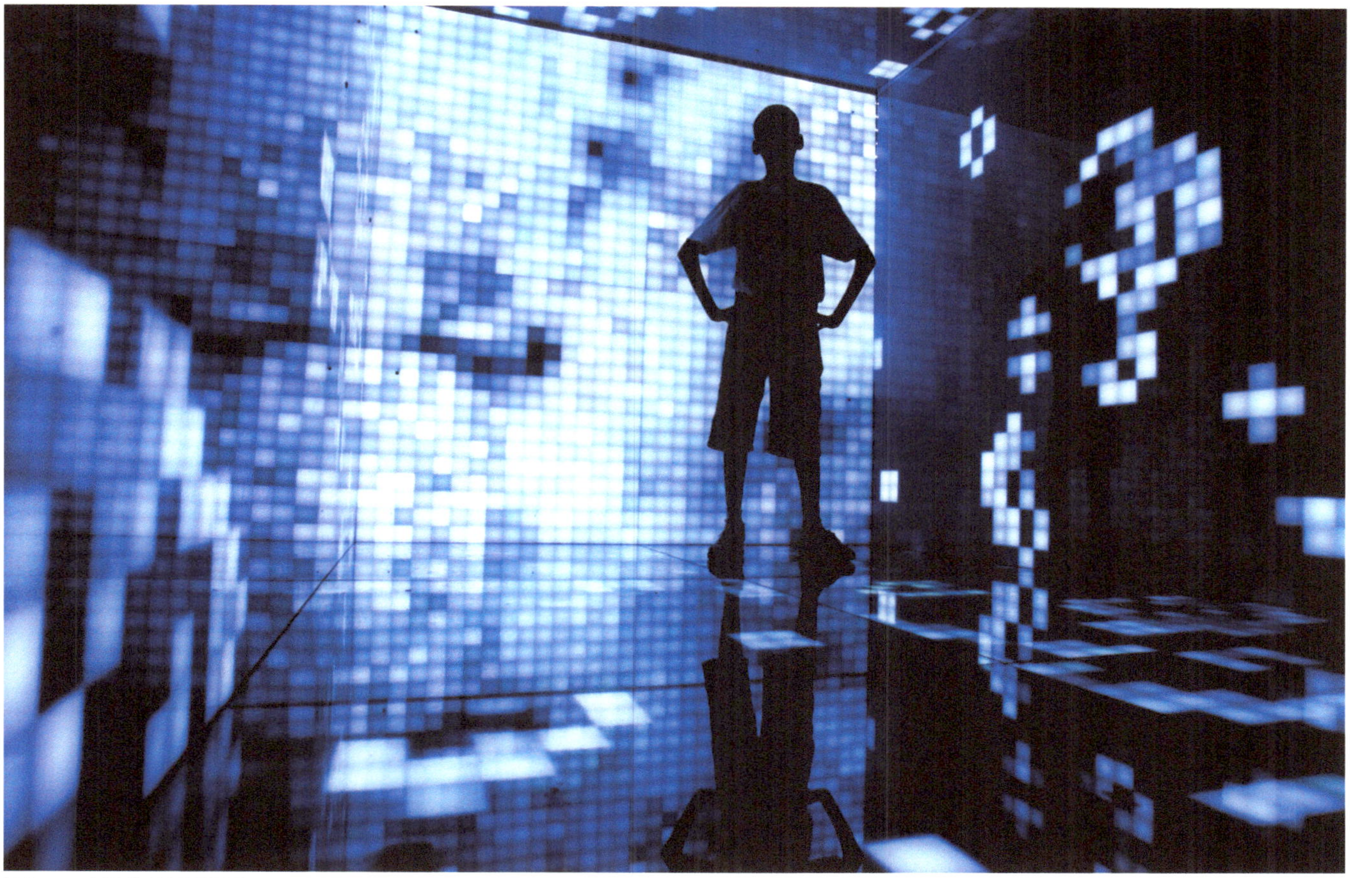

164. etoy, *MISSION ETERNITY*, 2007-ongoing

Design | Optimisation | Fabrication

Wendy Fok

As an artist, how do you see your role in a technological or scientific setting?

Architectural design today could possibly be described as the relational equations mediated by digital techniques assisted with production and knowledge of fabrication. Like many fields in the modern culture, it strives to be truly integrated where the designer can move seamlessly from concept to production in a single, contained process.

Reiser+Umemoto in "Atlas of Novel Tectonics" gives a pointed statement that the "Material practice is the shift from asking "what does this mean?" to "what does this do?" This crucial comment questions the genesis from interpretation and meaning to performance in architectural thought, moving beyond the instrumental notion of the traditional critiques that operate within metric spaces.

Born into the day when TIME magazine issued its "Computer Generation—a New Breed of Whiz Kids" cover, and into the age and generation where digital architectural design has the ability to fascinate through the abundant and superfluous availability of insatiable tools that are readily available to anyone—by means of "anyone", anyone is someone who has the ability to use a mouse. This poses a question on how our visions of objectivity diverge into the tendency on pushing and understanding the limits of different material properties to further the development of architectural design. As an augmented quantity of designers incorporate diversified parameters into their projects, an increased number of potentials within formal creation emerge; however, as an increase in the techniques of application become available, the lineage of sequential development within creation of the final product ceases to be coherently maintained. Thus, the fundamentality of mapping a layered approach within design intelligence between parameters and constraints simply becomes the decision and knowledge accumulation incorporated through organizational efforts disposed by the architect.

As discussions continue to propose the difficulty of utilizing the computer as an active part of a design process, designers begin to further the ability to negotiate between quick intuitive studies and definitive quantifiable decisions. These interpretive models then alter the design processes which the workflow of ideas can remain separate and abstract. As relationships between the parts in design coalesce, a new type of workspace is formed which deals primarily with associations of the parts and their interdependent relationship with the whole.

Concurrently, projects will require exploring the rigorous design methodologies that trace the developments of complex and dynamic forms in the context of their structural feasibility. While, the evolutions of topologies are of special interest, typological properties intrinsic within the technique of design will also be placed in consideration. Thus, in order to understand of "what does this do?", the practice of architectural design, surrounded by

investigative and re-combinative form finding processes and evaluation, should entail sustained lab testing of physical models within the systematic development of understanding both 2D and 3D modelling, in conjunction with computer simulation and the optimisation process.

Intermediary: The Scientific Evangelist

Jonathan Minard
Michael Pisano

A communicator who nimbly traverses domains of thinking, the Intermediary serves a necessary role by synthesizing, articulating, and promoting the importance of scientific ideas. They call attention to that which is yet unknown, and the greater implications of future discoveries: "As the circle of light [scientific knowledge] increases, the circumference of darkness also increases." (Albert Einstein).

This submission was written by Jonathan Minard and Michael Pisano, who were observers of the book sprint. The two noted the omission of the "intermediary" figure, and took it upon themselves to contribute this astute essay on the subject.
—Andrea Grover

In a culture fraught with narrow specialization and exclusive jargon, dedicated evangelists and interlocutors are necessary to translate emerging scientific knowledge between discrete communities. These Intermediaries function as high-level interpreters, literate in cross-disciplinary domains of expression and capable of clear communication of concepts and their relevance. In turn, this discourse awakens the public to the sense of wonder at the core of scientific inquiry.

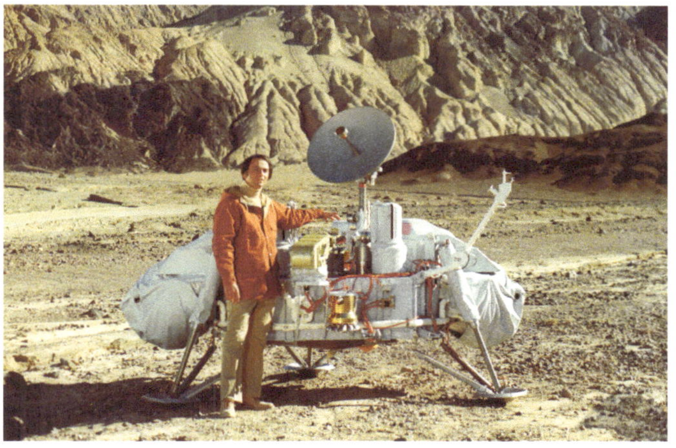

165. Carl Sagan, *Cosmos:* A Personal Voyage, 1989

Some of the most familiar examples of Intermediaries include popular scientists: Carl Sagan and Ann Druyan, Rachel Carson, E.O. Wilson, Brian Greene, and Neil deGrasse Tyson, who began their careers in research, later rising to prominence as public educators, writers, and media producers. For audiences, their inspirational messages have brought a new framework of meaning to Big Questions: the nature of time and space, our place in the universe, and the fate of all life on Earth. Before Sagan, Harvard astronomer Harlow Shapley advocated for a new "Cosmic theology" based on an earnest, modest kind of rationality: "We do not amount to much in size, or in duration either, for that matter: but we have the gift, I hope, of humility and reverence and we have an inborn impulse to learn and understand."

Intermediaries and their platforms bring a humanistic perspective to a given subject, contextualizing a granular level of understanding within a holistic picture of social, philosophical and spiritual concerns. Motivated by awe and curiosity at the edges of the known universe, these explorers translate otherwise invisible or inaccessible phenomena for public audiences.

An Intermediary may identify as a scientist, artist, entrepreneur, writer or filmmaker. A common drive is the impulse to relate to others through sharing of ideas, serving as conduits for the flow of information. They are storytellers, translators and emissaries capable of distilling complex concepts into an accessible narrative or poetic form. While they may be versed in the language of science, above all, they must be able to translate the rarefied vocabularies of narrow disciplines into general terminology.

In recent years, the infrastructure for disseminating information has become dauntingly dynamic. Media is now omnidirectional, interactive, navigated as much by its consumers as by its creators. Gone are the days in which a simple book or television program might reach a vast audience: today, the constant growth and diversification of our vast multimedia palette calls for a new hierarchy of specialized practitioners. The volume of accessible content is equally unbounded, allowing for even well-articulated information to remain obscure without savvy handling. In suit, some Intermediaries are active in the creation and management of new platforms for dialogue, aware of shifting attention with respect to new media.

Personal Narratives

The approach of the Intermediary is necessarily empathetic—they are participant observers able to reconcile divergent points of view and internalize the perspective of others. By treating the subject at hand with a measure of warmth and humanity (pathos), they help us become comfortable with the alien and unfamiliar. Jane Goodall's work with primates and larger conservation issues is a clear example of humanizing the foreign; closer to home, physician, neurologist, and writer Oliver Sacks adopts a similar approach. As a portrait-artist of neurological anomalies, he gives his readers insights into the mind's architecture by allowing them to momentarily inhabit a strikingly divergent way of being. Combining his powers of inference from a life of clinical work with a profound degree of compassion, Sacks gives voice to patients whose alienating disorders render them incapable of relating to others or defining their own condition–be it prosopagnosia (the inability to recognize faces), uniquely gifted savants or people with acute amnesia. Sacks remarks:

For me, an interest in science is inseparable from... storytelling. In medicine, of course, narratives are essential: the patient tells you what's going on, and you try to match this with stories heard from other patients. I love to give personal accounts, to try and enter people's experiences and describe them.
(Interview, Universe, Claire L. Evans).

E. O. Wilson's *Naturalist* (1994, Island Press) weaves scientific discovery into an engaging autobiography, providing a lyrical point of entry into a specialized field. Similarly, *Journey to the Ants: A Story of Scientific Exploration* (Wilson and Bert Holldober, 1994) reframes modern myrmecology (the study of ants) in accessible language and personal narrative, allowing for discipline-specific conclusions to provide broader sociobiological implications. Facts and data from field work are co-opted as semi-formal storytelling devices, both in breaking up the novel's structure and to add a parallel narrative (at times, bordering on extended metaphor) to the personal experience of a scientist.

In general, Wilson's promotion (and, in some cases, coining) of neologisms such as "biophilia," "consilience," "scientific humanism," and the "evolutionary epic" effectively bridges scholarly pursuits with the deep humanistic conclusions he draws from specialized research. Consilience in particular, which he defines as, "Literally a 'jumping together' of knowledge by the linking of facts and fact-based theory across disciplines to create a common groundwork of explanation" (p. 7, *Consilience: A Unity of Knowledge*). This demonstrates Wilson's strengths as an Intermediary: personal passions for his specialization within natural science are easily matched, if not surpassed, by a commitment to foster collaboration between the sciences, humanities, and the public.

To the extent that each person can feel like a naturalist, the old excitement of the untrammeled world will be regained. I offer this as a formula of reenchantment to invigorate poetry and myth: mysterious and little known organisms live within walking distance of where you sit. Splendor awaits in minute proportions. (Wilson, Biophilia, 1984, p. 139).

In the genre of science writing, non-specialists perform an equally important role by exploring human interest topics beyond a single area of research. A naive outsider may take the approach of an anthropologist, entering a scientific community as a participant observer. For a recent book, the popular science writer Mary Roach took a ride on NASA's vomit comet to float in zero gravity and experience the nausea and ecstasy of weightlessness. Roach, who became famous for STIFF, a work on the post-mortem careers of human cadavers, follows her infectious curiosity on adventures into worlds hidden from view: most recently, sneaking through the back door of space agencies, from NASA and Japan's JAXA, and recording the lurid details of astronauts' life and work.

Packing for Mars; Curious Science of Life in the Void (2010) takes an uncensored approach to science journalism, addressing common questions about human behavior in space which government agencies do not readily disclose. Through her exhaustive recording of facts, she paints a picture of the human space program that balances the glorified image of astronauts with the tedious reality of their jobs, detailing the strenuous psychological and physical trials they must endure. With the candor of a health-ed teacher, she explains how astronauts manage every imaginable bodily function in zero gravity: from sleep, diet, and exercise to defecation, vomiting, and sexual activity. No unpleasant descriptions or prurient curiosities are omitted. Mary Roach's readers benefit from her stance as a sophisticated layman, an outsider in the worlds she honestly reports from.

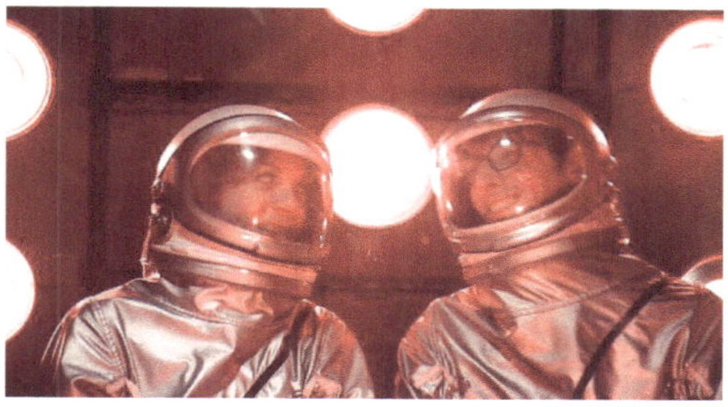

166. Still from *Packing for Mars: Space Hygiene*, Mary Roach (2010)

Establishing Forums for Dialogue

Intermediaries may take a purely interrogative approach, boldly posing the questions that scientists closely focused on the inquiry at hand may not stop to consider, or those completely removed from the conversation may not be equipped to ask. John Brockman, the publisher and founder of Edge Foundation, has made a career generating salient questions. He promotes a "third culture" to bridge the chasm of science and humanities, and nurture the emergence of a "new natural philosophy, new ways of understanding physical systems, new ways of thinking about thinking that call into question many of our basic assumptions" (Brockman, what we believe but cannot prove, x). Each year, Edge poses a single question to a sampling of the hundred or so "most interesting minds in the world" and anthologizes the resulting mini-essays from luminaries such as Paul Davies, Kathyrine Denning, and Stephen Pinker. Questions can be open-ended and far-reaching as: "What do you believe but cannot prove?", "What will change everything?", and "Is the Internet changing the way you think?" The nature of these collections encourages the contributors to engage in reflection, outside the constraints of scientific method, and permits speculation free from the scrutiny of academic peer-review. Such a forum favors open-ended intellectual discourse with the potential for sparking debate where contradicting opinions collide. Science is a dialectic in which no topic is ever fully settled and everything remains an open question.

Scientists and their intellectual allies are looking beyond their individual fields—still engaged in their own areas of interest but, more important, thinking deeply about new understandings of the limits of human knowledge. They are

seeing our science and technology not just as a matter of knowing things, but as means of tuning into the deeper questions.
(Brockman, *What We Believe But Cannot Prove*, 2006, xii)

Working against close-mindedness and apathy, Intermediaries bridge cultural divides by advocating for, and evangelizing on behalf of, ideas. Most importantly, they present the relevance of concepts beyond the narrow scope of everyday life, by expanding the frame of reference to a bigger here, and longer now. Stewart Brand—big thinking advocate for the environment, and initiator of global dialogues about the future—asserts "Science is the only news. Human nature doesn't change much... Science does, and the change accrues, altering the world irreversibly."

Long Now Foundation, established in 1996, is dedicated to moving the cultural mind-set away from myopic, short-term thinking to consider the future beyond our lifetimes, attempting to "[do] for thinking about time what the first photos of the Earth from space have done for thinking about the environment." (www. long now.org). Seminars About Long Term Thinking (SALT), similar to Edge.org and TED talks, facilitate a multilogue among astute, forward thinking minds. Founders Brand, Kevin Kelly, and Brian Eno moderate discussions and audience questions. The seminars encapsulate the foundation's broad purview, which is oriented around a 10,000 year clock project, addressing concepts such as time and responsibility, long term storage of information, and massive sociological trends (poverty, population growth, cities, technology, energy use, etc.). The discourse focuses on anthropogenic change, and prospective strategies to remedy these mounting issues. SALT takes place once a month in the Bay Area, reaching a global audience through online distribution of audio podcasts.

Edutainment

The production, presentation and distribution of educational content for broadcast media like television relies on a hybridized tone to co-exist (and compete) with other serialized entertainment. Intermediaries have been successful in using popular entertainment infrastructure (and related channels of re-distribution, from DVD release to Hulu) as a tool for outreach by flavoring exposition with humor, drama, action, and other cues from non-educational productions. Popular Science shows such as Mythbusters infuse thematics of danger and challenge, as well as fast-paced, segmented editing practices from other pop culture genres, into technical inquiry. Some Intermediaries, including Carl Sagan, Bill Nye, and David Pogue, act as hosts or guides capable of hooking viewers through personality and performativity; others, such as *How It's Made* or *Shark Week*, focus on publicly inaccessible subjects, and deliver captivating imagery. BBC's *Planet Earth* benefits from both sublime exclusive content, set to an entrancing narration by David Attenborough.

The public nature of broadcast media as a platform allows for a certain amount of evangelism alongside education, both directly and by creating inherent value of scientific progress as the subject of a consumable commodity.

People are not going to care about animal conservation unless they think that animals are worthwhile...[they] must feel that the natural world is important and valuable and beautiful and wonderful and an amazement and a pleasure.
(Attenborough)

In addition, the potential revenue from high-quality broadcast productions allow for use of new technologies—from the early adoption of digital cinema quality and 3D video hardware, to images from the Hubble telescope. (At the time of this publication, Discovery and Xfinity have, in the last few months, unveiled the first 3D HD channels, featuring a slew of science and nature programs).

Poetic Documentary

A generation before Jacques Cousteau explored marine landscapes in *Silent World* (1956), filmmaker Jean Painleve invented one of the first underwater housings to film aquatic life, and embarked on an prolific career obsessively documenting animal behavior. From *The Love Life of the Octopus* and *Underwater Assasins* to *The Vampire (Bat)*, Painleve's lens captures the strangeness of familiar organisms with an often titillating focus. These films not only anticipate the genre of nature documentaries, but subvert scientific objectivity in favor of a more interpretive and experimental approach. His soundtracks often featured free-jazz or psychedelic instrumentals, amplifying the films' weirdness. The recently released anthology of Painleve's work, titled *Science is Fiction,* celebrates theradical poetry and surrealism characteristic of his vision. Compared to the contemporary nature documentary genre Painleve's work incepted, his films still feel strikingly subversive.

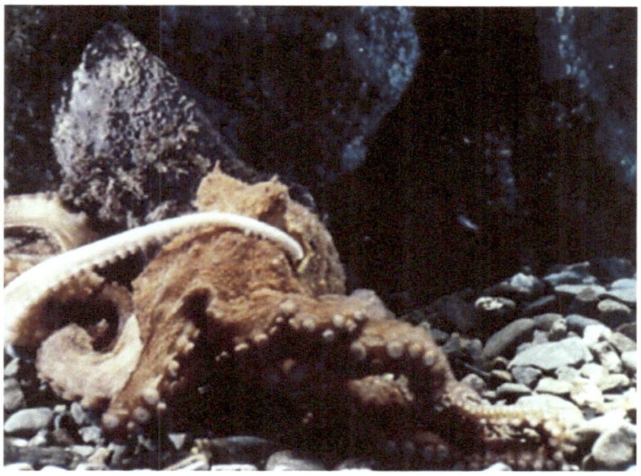

167. Still from *The Love Life of the Octopus (Les amours de la pieuvre)*, Jean Painleve (1965)

Poetic documentary distills a non-fiction subject through the lens of a strong voice, imbuing content with emotion and distinctive point of view. As opposed to following a traditional story arc, the structure and ambience of the presentation strikes a tone, and builds resonance on ideas, experimenting with perceptual shifts and discontinuities. Application of special effects, in-camera or post-production techniques, may achieve a heightened auditory or visual experience related to the content. Practitioners work in a more open-ended, free-form and ambiguous field of both ideas and technical structures (i.e. experimental film and video art), requiring the viewer to interpret the content from an unfamiliar level of abstraction.

In *Powers of Ten*, a 1968 film by Charles and Ray Eames, the camera telescopes out from a human scale of perception, climbing the orders of magnitude from the atoms in a human hand to the galaxies and back (all in under ten minutes). By comparing the size of an atom to a human hand (and all the trillions of molecules contained within it), microscopic space is shown to be just as vast as the universe in relation to our planet. Contextualizing a human point of view within a holistic framework displays the interconnectedness of each point along the spectrum. IBM sponsored the creation of the film as an instructional tool, to provide a visual aid to teach students about relative scale; its legacy extends beyond that purpose, as a widely-referenced masterpiece of experimental film.

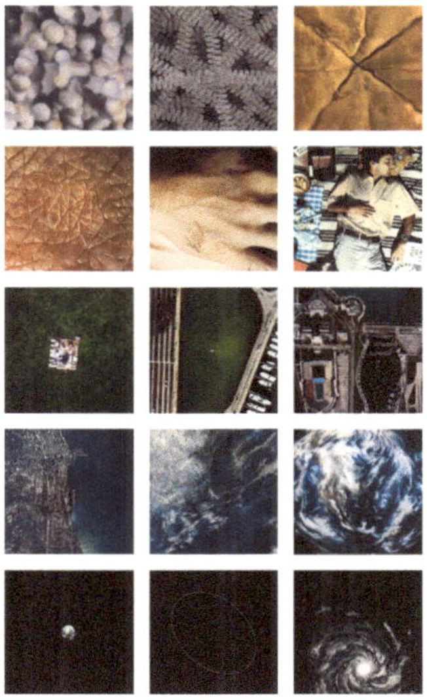

168. *Powers of Ten*, Charles and Ray Eames (1968)

Kerry Tribe's *H.M.* (2009) is "a two channel film [installation] based on the true story of an anonymous, memory impaired man, known in scientific literature only as 'Patient-H.M.' " The patient's acute amnesia resulted from an operation removing a part of his brain, at the source of severe epileptic seizures. The filmmaker used an actor to restage interviews with H.M. based on archival film of the actual patient. Its form attempts to shift perception by immersing the viewer in a jarringly disparate experience: "A single 16mm film plays through two synchronized projectors with a 20 second delay between them. The structure of the installation and the nature of the material together produce a sensation of mnemonic dissonance much like that experienced by patient H.M." (Tribe). By attempting to replicate the condition through the experience of watching the film, Tribe literally allows the viewer to inhabit the subject's reality (much like Oliver Sacks strives to achieve through empathetic storytelling).

Biography

Character studies, independent of medium or platform, lend a strong anchor point for an audience that might otherwise be distanced from abstract or unfamiliar concepts. Errol Morris, a documentary filmmaker known for probing interviews with highly eccentric individuals, invented a camera system called the InterroTron; a machine that uses a teleprompter to project a live-video image of the filmmaker's face in front of the camera's lens, thereby engaging subjects in direct eye contact. The resulting intensity of gaze is passed to the viewer, forcing confrontation with a stranger's foreign ideas. In *Fast, Cheap & Out of Control*, roboticist Rodney Brooks explains how the study of insects inspired him to think about artificial intelligence and robotics in evolutionary terms—leading to the creation of machines that explore the world using simple feedback circuits embedded in their limbs. Morris' series of interviews at IBM Labs for an unfinished documentary, *True Strangeness of the Universe*, captures computer scientists waxing poetic about math and logic. Other interview subjects have included Clyde Roper (Giant Squid hunter), Gretchen Worden (Director of the Mutter Museum), Saul Kent (Founder of the Life Extension Foundation), and Temple Grandin (animal behaviorist/savant).

Much of Morris' work examines the minds of fringe scientists, people who straddle the line between genius and quackery. A dramatic tension is created between our inherent skepticism and empathy: rational response to exotic viewpoints comes up against a desire to suspend disbelief for the sake of another. It is common for biographical documentaries to take on obscure or misunderstood characters: *Heaven + Earth + Joe Davis*, Peter Sasowsky's documentary about M.I.T. biologist-artist Joe Davis, paints an endearing portrait of a visionary on the fringe, lending credibility to a character who might easily be portrayed as insane without a sympathetic filmmaker.

Strange Culture (Lynn Hershman Leeson, 2007) is a dramatized account of biologist-artist Steve Kurtz, who was wrongfully accused of being a bioterrorist following the death of his wife, and subsequent search of his studio. Through their portrayal of the FBI's confiscation of his basement lab, his art practice with collective Critical Art Ensemble (CAE), and a prolonged trial, the filmmakers elevate Kurtz to heroic status. CAE itself acts as an Intermediary, using tactical media to address political and social issues at the intersection of science and culture. In particular, they focus on genetics and bioengineering, raising public awareness of questionable corporate practices in contemporary food and drug industries.

Remixing and Post-Production

RadioLab, a nationally syndicated radio show on WNYC, creates unique broadcast journalism by layering sounds and stretching time to produce lush auditory textures. Their radical style of post-production evokes the formless chatter of thoughts and phantom images in the brain.

Each episode is a synthesis: a remix of stories, interviews, found audio, and electronically generated sounds that meld information. Hosted by Jad Abumrad, an electronic composer who designs the show's soundscapes, and Rob Krulwich, a science journalist, the duo casually banter their way through topics as broad and deep as "Language," "Stochasticity (Randomness)," "Love," and "Mortality." Radiolab uses fantastic metaphors and real-world examples to anchor pleasingly cerebral conversations with guest scientists and artists, stumbling upon moments of epiphany . In a single episode on the topic of "Time," they float through a recording of Beethoven's 9th symphony digitally stretched into a twenty four hour drone, an interpretation of Muybridge's high-speed sequential photography of a horse-race, and a poignant audio time-lapse (twelve years condensed into 4 minutes) of a child's speech patterns developing from baby babble to complex sentences.

As defined by art theorist and curator Nicholas Bourriaud, post-production extends beyond the practice of authoring original content for film, video, television, or radio. This expanded definition of Post-Production refers to the repurposing, remixing, and re-cutting of existing artworks into a new form through an alchemical creative process, of fusing content from multiple sources.

[The] art of postproduction seems to respond to the proliferating chaos of global culture in the information age... These artists who insert their own work into that of others contribute to the eradication of the traditional distinction between production and consumption, creation and copy, readymade and original. (Post-Production, Bourriaud).

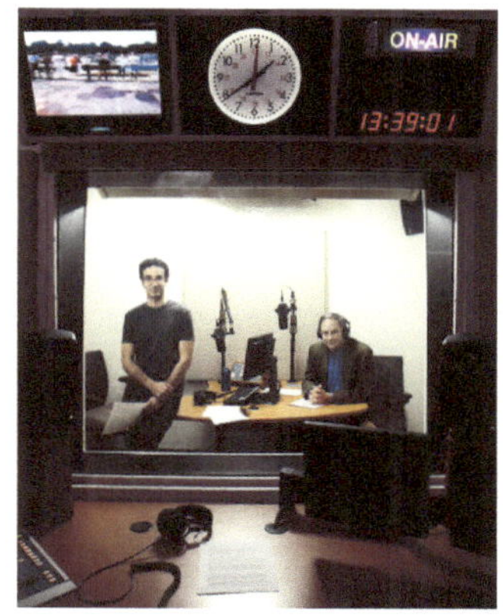

169. Jad Abumrad and Rob Krulwhich, hosts of RadioLab WNYC

While many traditional media productions gather from a breadth of primary source information to catalog, synthesize, and present concepts, the rise of digital media creates new modes of expression and promotion: Intermediaries in this domain appropriate, recut, and remix those secondary sources to make something entirely new through the process of post-production. This borrowing and recontextualizing of source material enhances or augments its meaning, and simultaneously extends the lifespan of ideas through new iterations and popular formats. The relative ease and democratization of digital production also creates a dialogue between practitioners, spurring rapid responses and further re-purposing of source material.

Virality

Between the Voyager Record, Pioneer Plaque, and Arecibo message, Carl Sagan's ideas have traveled the greatest physical distance of any Earth-made product. His reach into the universe may surpass any other human's. Today, Sagan's posthumous outreach is similarly broad in the digital repetiton and manipulation of his messages. Wide re-distribution kept him in the public consciousness post-syndication of *Cosmos: A Personal Journey* (produced with Ann Druyan in 1980), one of the most popular science television shows of its time. In cyberspace, the pervasion of his ideas is equally vast: drawing content from Sagan's series and specials, a contemporary set of digital Intermediaries remix and re-present his ideas through Youtube and other online channels. In his second life on the web, Sagan's messages maintain a massive cult following, promoting Sagan himself to an icon of secular humanism. Here, the role of the Intermediary has been to recognize, and adapt with, new media concepts such as virality and memes, in order to update relevant content.

Some of the most contagious ideological memes at the essence of Sagan's message: the vastness of space, wonder of science, the fragility of humanity, our great responsibility in the face of our destructive capability, the mediocrity principle (that life must be pervasive in the cosmos), and the celebration of our diversity and common heritage as residents of "a pale blue dot" in the "backwaters of our galaxy." For many viewers, it is enough to watch original source material vetting these thoughts online; others are captured through artful remixing, and viral formats in which culture references culture.

NASA has glommed onto the trend of audio-visually remixed viral videos, as displayed in Parts I and II of *The Sagan Series*, their own high-production Youtube tributes to Carl Sagan (receiving approximately a million hits, produced for NASA by Reid Gower, with music by Michael Marantz). The voice of a beatified Sagan soliloquizes over aerial sweeps of Earth. A solemn tone struck by a piano elegy gradually grows hopeful, as visuals move from juxtapositions of fragile nature, burgeoning technology and human strife to celestial horizons. It is a brilliant marketing strategy for NASA, which has recognized that Sagan's messages are a persistent meme steadily propagating through the web. By linking Sagan to NASA, his inspiring message can benefit the image of the federal space program.

Symphony of Science (www.symphonyofscience.org) hosts a series of video remixes, sampling popular cosmologists—Carl Sagan, Stephen Hawking, Richard Feynman, Neil Degrasse Tyson, and even Bill Nye—and musically augments their voices with vocoders (auto-tuning a melody to mimic singing) and analog synths. Striking a tone of cosmic campiness, the series riffs on these scientists as hybrids of new age prophets and pop stars. Hilariousness aside, the videos serve an important function for the subjects they playfully represent, freely proliferating their ideas and accomplishing creator John Boswell's goal of delivering science and philosophy to the masses (*We Are All Connected*, feat. Sagan, Feynman, Tyson, and Nye, garnered over 3 million hits). In part, proliferation of viral media led to a general need for high production value in online video, especially for organizations seeking to simultaneously exploit social distribution channels and remain credible.

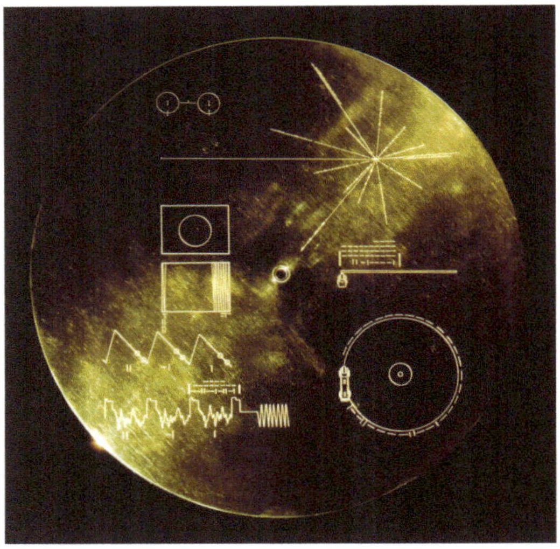

170. "Sounds of Earth" gold-plated record and U.S. flag prepared for storage aboard Voyager 2

Online Distribution

Part of understanding and exploiting online infrastructure as an Intermediary is the creation of new media forums around the technology's rapidly progressing capabilities. Creating a platform for presentations sets out to spread ideas, and focus the attention of an online audience somewhere with meaningful content. The TED Talks, curated by Chris Anderson, have established one of the most watched online forums (over 290 million online views as of July 2010) for interdisciplinary dialogue. Not only the videos themselves boast a high production value, but the presentation tools within the talks dazzle viewers with gorgeously designed motion graphics. Statistician Hans Rosling's presentation, "The best stats you've ever seen" (which received over 7 million views on Youtube) features animated graphs and an energetic delivery to illustrate global trends in socioeconomic progress.

A great TED Talk delivers what every PowerPoint presentation aspires to: intelligent design on par with the staggering graphs and animations in Al Gore's *An Inconvenient Truth* (2006, Directed by Davis Guggenheim). In recent years, the rising status of the talks and the daunting precedent set by inspirational lectures—such as Jill Bolte Taylor's "Stroke of Insight"—have led to a pattern of one-up-manship in oratory and competitive performance. The format, an 18 minute talk, seems tailor-fit to our modern attention spans, providing just the right amount of time to squeeze in a fast-paced, yet substantive, presentation. TED provides often under-recognized, but highly inventive or visionary, thinkers with a platform to market their ideas in one of the web's hippest venues for nerds. Through their marketing and branding, TED assumes an epic identity, offering online audiences the chance to be uplifted, enlightened, and entertained by skillful presentations.

If manipulated correctly, the inspirational effect of these communicators on the general public can have a galvanizing impact. This power is being directed to encourage people to take action in the interest of large-scale change. "TED Prizes" are granted to a handful of recipients to design and promote their "One wish to change the World." In 2011, Jill Tarter of the SETI institute encouraged everyone to join in the search for "cosmic company."

In 2007, E.O. Wilson presented his wish "on behalf of all creatures, that we learn more about our biosphere and build a networked encyclopedia of all the world's knowledge about life".

The latter has resulted in the Encyclopedia of Life (EOL), a crowdsourced database with the goal of creating a single editable webpage for every species on Earth. It attempts to aggregate and centralize information, pooling public and specialized resources towards a common goal: to gather data and promote awareness about the need for conservation. It's not just a scientific effort. It has a strong message about why biodiversity is important for humanity and the planet.

Chris Anderson notes that Wilson's EOL exemplifies the kind of large-scale collaborative projects that the TED Prize was designed to facilitate. "A lot of people had been thinking about this idea based on his work for years, and hadn't been able to make it real. And so you had this huge surge of support for him." (2006 conversation with Charlie Rose).

Scientific Evangelism

An ambition of Intermediaries is to establish points of entry into exclusive pools of knowledge. Prevalent strategies include drawing connections to a subject through personal narrative (as with E.O. Wilson's *Naturalist*), and causing shifts in perspective that are long lasting and broadly applicable (as with *Powers of Ten* and the Long Now Foundation). The abilities to distill enormous ideas, and to relate the particulars of a specialized field of knowledge to the larger whole combine to foster holistic thinking. Finding alternatives to reductionist thinking and statistical simplification, the Intermediary can add benefits to the scientific method through a high tolerance for nuance, complexity, and paradox learned from the arts and humanities.

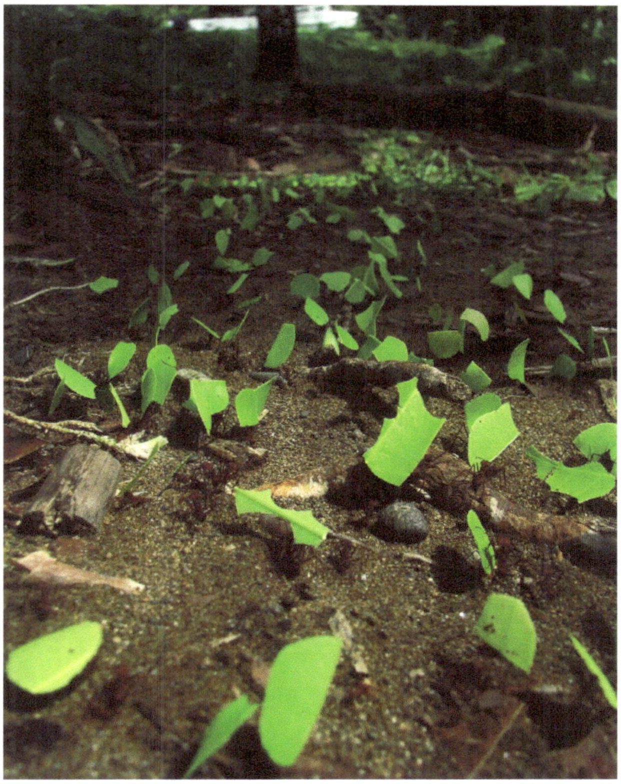

171. E.O. Wilson's Leafcutter Ants

Scientific evangelists communicate the importance of research both across disciplines, and to a wary public. Missing pieces and contested theories can be cast as opportunities for further expansion of consciousness and catalysts for innovation. This shift in tone not only encourages cross-disciplinary thinking as a tactic, but also facilitates it by making content accessible to diverse practitioners and lay audiences.

Intermediaries combat impediments to dialogue: skepticism, apathy and naivete. Common misconceptions, such as the external view of science as absolute, preclude audiences from realizing the prevalence of the unknown and unmeasurable in science. A true translator questions this understanding as an integral narrative component. One reason for the convergence of science and the humanities is that there is a limit to what can be tested and verified. For example, the notion posited by string theory that there might be 11 dimensions of space-time is purely abstract, a model arrived at through mathematics (as opposed to an experimental method).

Science will never empirically verify what was here before the universe came to be, and what will result from its inevitable expansion. In certain fields, researchers have reached a critical threshold of testability, a transition point from scientific inquiry to philosophy, to metaphysics. Approaching science with an epistemically critical perspective, Intermediaries create space for and moderate higher-level discussion around the nature of knowing, and questions about how we know what we know. Why, beyond practical application, is humanity engaged in science? One answer is for a deeper understanding of ourselves: of how humans relate to the universe and the scientific study of it, and how we can relate to each other at a time when specialization hinders discourse.

The challenge to maintain and articulate this attitude as germane to contemporary dialogue calls for specialized Intermediaries. These are the creators of new media platforms, and the practitioners who co-opt new infrastructures as communications tools. Evangelism in the context of rapidly changing social technologies and trends almost seems to require a forward-thinking, visionary approach to producing and leveraging creative assets: from Jean Painleve's technical innovations and association with the French Surrealists, to Radiolab's avant-garde podcasts and Sagan-inspired Internet memes, there is a clear radical thread linking Intermediaries in all their incarnations.

There is both an art and a science to creating understanding, and as a result, those who seek to create it reflect a hybridized, often intuitive, scholarship: diplomacy, pedagogy, and humanism need to be cross-bred with professional creative practice and technological savvy, either embodied in an exceptional individual or a committed team. Any of the above qualities can stand alone to engender awareness, interest, and dialogue; however, it takes Intermediaries with all of these qualities to inspire the next generation of open-minded thinkers.

JONATHAN MINARD is an artist and filmmaker who investigates human experience in extreme environments and the evolutionary dynamic between nature and culture. Recent documentaries have featured communities of extremophiles; the nomads of Mongolia, deep sea oceanographers, and artists working in outer space. Jonathan works as a director and writer for deepspeed media, currently in production on a documentary about the first privately funded mission to the Moon.

MICHAEL PISANO is an artist and writer in Pittsburgh. He works as a Writer, Producer and Editor at deepspeed media, currently on a documentary about the first-ever privately funded Moon mission. Michael aspires to a lifelong creative practice as an Intermediary.

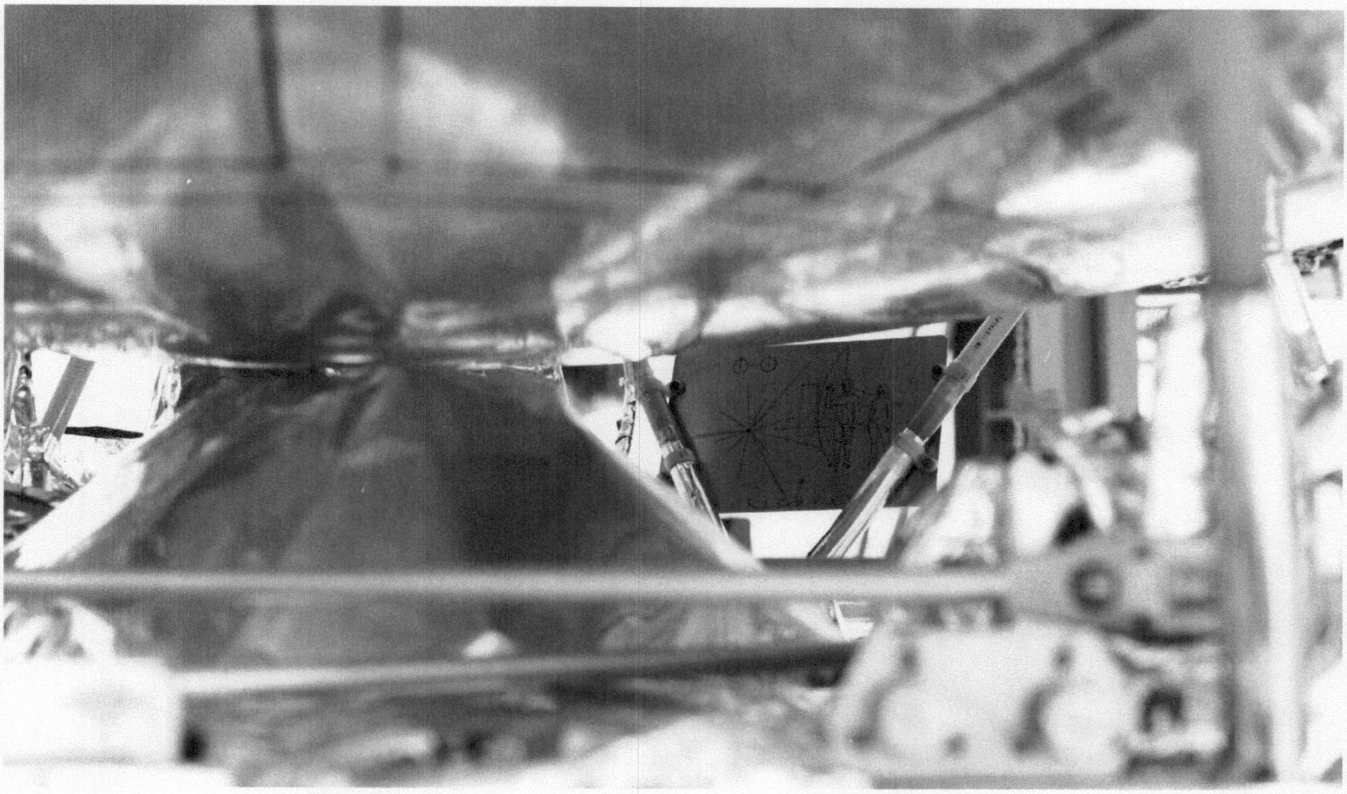

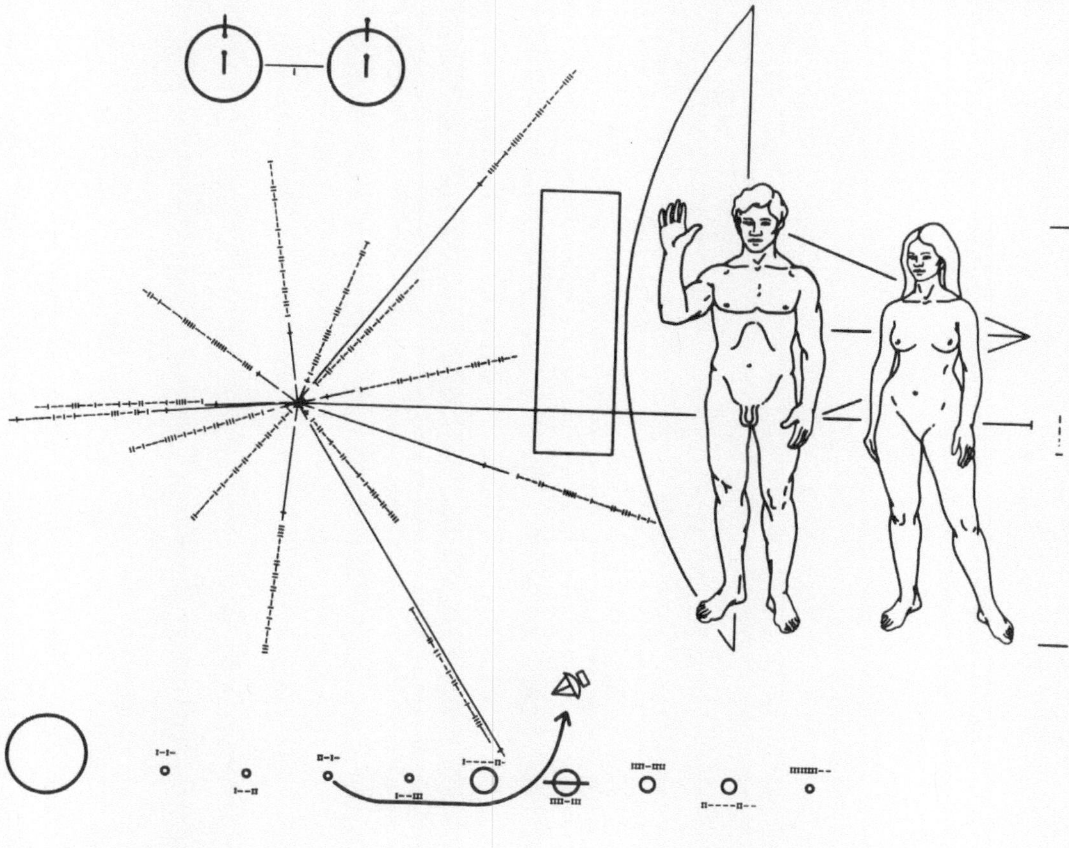

172. Pioneer Plaque, 1972

A SUBJECTIVE TIMELINE OF ART, SCIENCE & TECHNOLOGY INTERSECTIONS

WHILE WE USED REASONABLE EFFORT TO INCLUDE ACCURATE INFORMATION IN THIS TIMELINE, WE MAKE NO GUARANTEES THAT IT IS FREE FROM ERROR.

1637
— *Discourse on the Method of Rightly Conducting One's Reason and of Seeking Truth in the Sciences*, a philosophical and autobiographical treatise, is published by René Descartes in Leiden, Netherlands. It is the source of the famous quotation "I think, therefore I am." According to Lawrence Weschler at the 2011 *Art as a Way of Knowing* Conference, this precipitated the fissure between art and science.

1727
— Junto debating society is formed by Benjamin Franklin. The group held weekly meetings to discuss moral, political, commercial and scientific topics of the day.

1748
— Julien Offray de La Mettrie publishes *L'homme Machine*, depicting humans as machines.

1760s
— Johann Schröpfer, a coffee shop owner in Leipzig, begins using magic lanterns to conduct séances, complete with projections of ghosts and spirits. Imitators such as Paul Philidor produce entire shows around magic lantern effects; by 1789, phantasmagori is a popular attraction.

1778
— Hannah Mather Crocker organizes a female reading society in Boston to study science and read the *Belles-lettres*.

1794
— The Conservatoire National des Arts et Métiers (Conservatory of Arts & Industry) is founded in Paris.

1818
— Mary Shelley publishes *Frankenstein, or Modern Prometheus*. 1

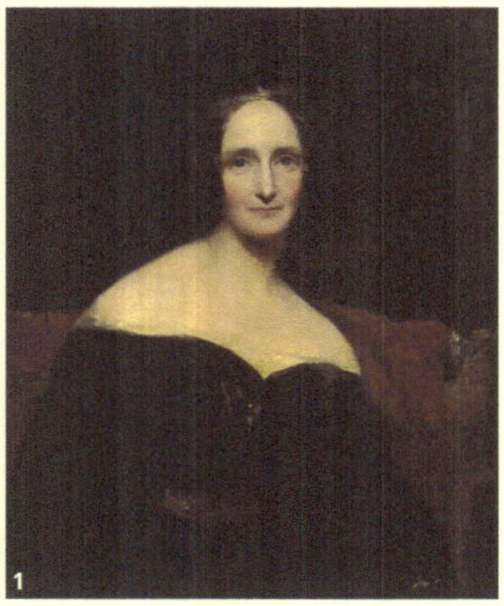
1

1823
— The first Mechanics Institutes are established in Glasgow and London, built on the foundations of a group started at the turn of the previous century by George Birkbeck, who instituted free lectures on arts, science and technical subjects. Such institutes spread worldwide, with the goal of self-improvement through education in science.

1826
— The Lyceum Movement is launched by Josiah Holbrook in Millbury, Massachusetts, created for "the advancement of education" and "the general diffusion of knowledge" via self-instruction and mutual improvement in all disciplines.

1826-1840
— Photography is invented: Niépce's heliograph, Fox Talbot's calotype, Daguerre's Daguerréotype.

1832-1837
— The electromagnetic telegraph is independently invented by European and American inventors.

1832-1834
— "Illusion toys" (phenakistoscope and zoetrope) are invented as early animation devices that produce an illusion of motion from a rapid succession of static pictures.

1833-1834
— The Analytical Engine is designed by Charles Babbage as a mechanical general-purpose computer.

1838
— Charles Wheatstone describes stereopsis, leading to the production of stereoscopes, devices for viewing three-dimensional images. 2

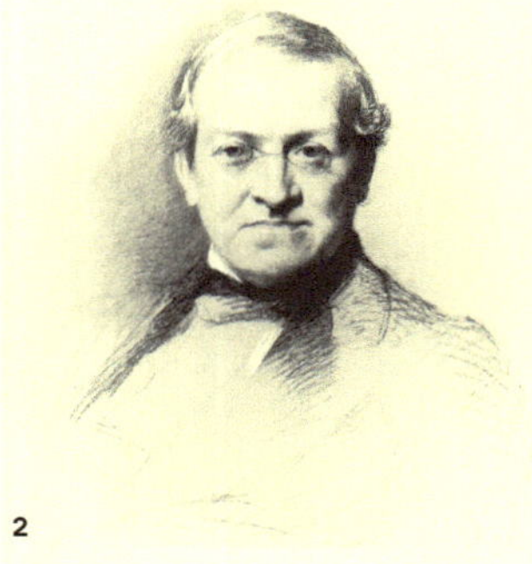
2

1838
— The first public demonstration of a telegraph is given by its inventor, Samuel Morse.

1843
— Ada Lovelace publishes her notes and a translation of L. P. Menabrea's paper on an algorithm encoded for processing by a machine, which is considered the first computer program.

1844
— Samuel Morse sends the first electronic telegraph in the Americas, transmitted via a repeater: *"What hath God wrought."*

1851
— The Great Exhibition in London is the first World's Fair, exhibiting "culture and industry". 3

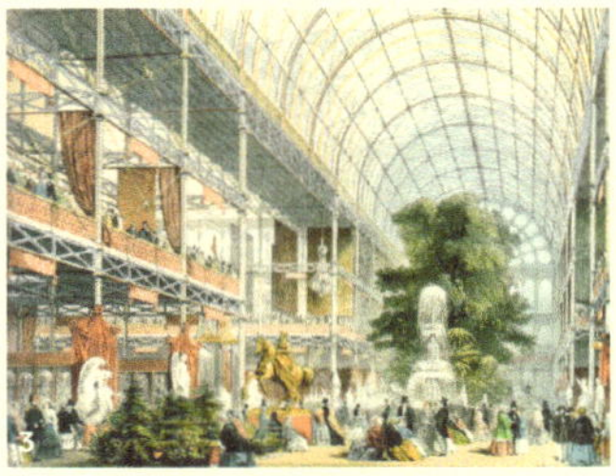

1852
— *Popular Science* magazine is launched by Edward L. Youmans to disseminate scientific knowledge to the educated layman. 4

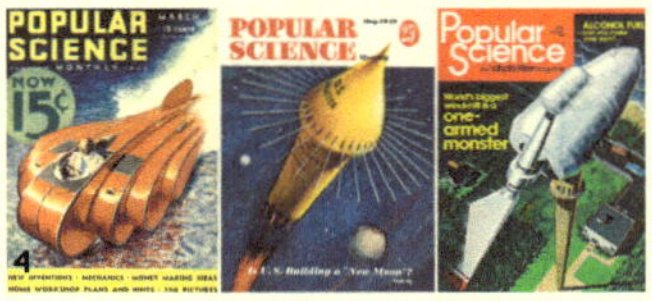

4

1863
— John Pepper adapts Henry Dircks' idea of manifesting an apparition through strategic lighting and an angled pane of glass for an on-stage special effect. The ghost appeared on stage during a scene, surprising the unaware public. Despite Dircks' innovation, the effect has been known as "Pepper's Ghost" ever since. 5

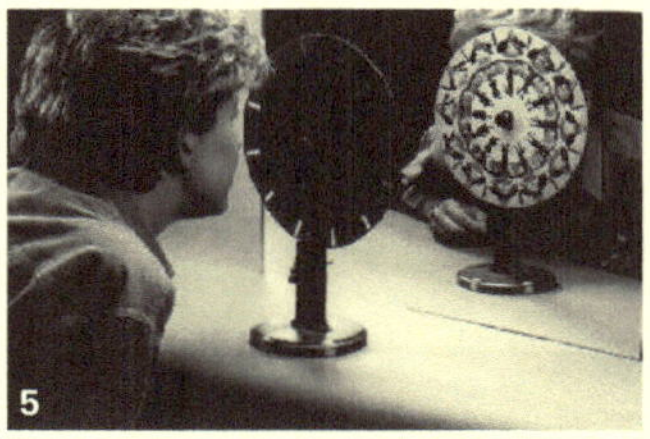
5

1876
— Alexander Graham Bell receives a patent for the telephone.

1887
— Eadweard Muybridge produces "Animal Locomotion," a piece of proto-cinema.

1888
— Fast, flexible film—*Kodak*—is produced by Eastman.

1893
— Thomas Edison and his assistant, W.K.L. Dickson, complete the Black Maria, America's first movie studio.

1895
— The Lumière Brothers hold their first public screening of projected motion pictures.

1896
— George Méliès invents the "stop trick," and pioneers other special effects, such as multiple exposures, hand-painted frames, time lapse, and dissolves.

1900s

1902
— H. H. Windsor launches *Popular Mechanics*, a weekly magazine devoted to science and technology. 6

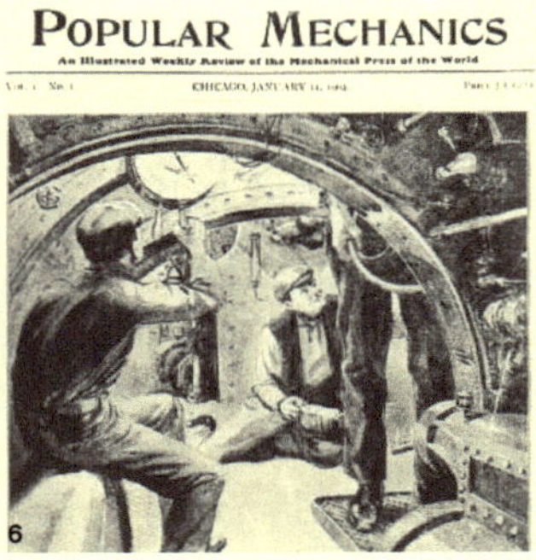

6

1905
— Albert Einstein publishes "On Electrodynamics of Moving Bodies," establishing his theory of Special Relativity.

1909
— Futurism, an artistic and social movement, originates in Italy. 7

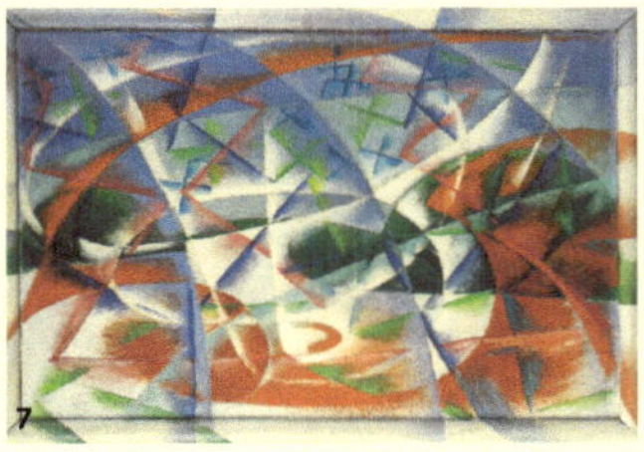
7

1915
— The first American coast-to-coast telephone call. 8

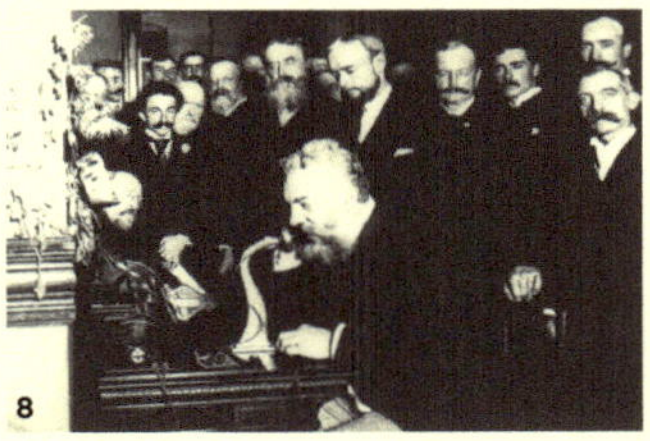
8

1919
— The Bauhaus school is established by Walter Gropius in Weimar, Germany. 9

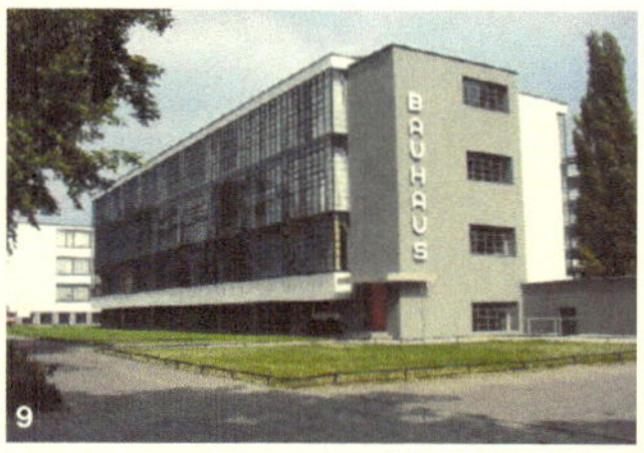

9

1919
— The Russian Constructivism movement begins, advocating art as a practice directed towards social purposes.

1920
— Léon Theremin invents the Theremin, the first "electronic" instrument. 10

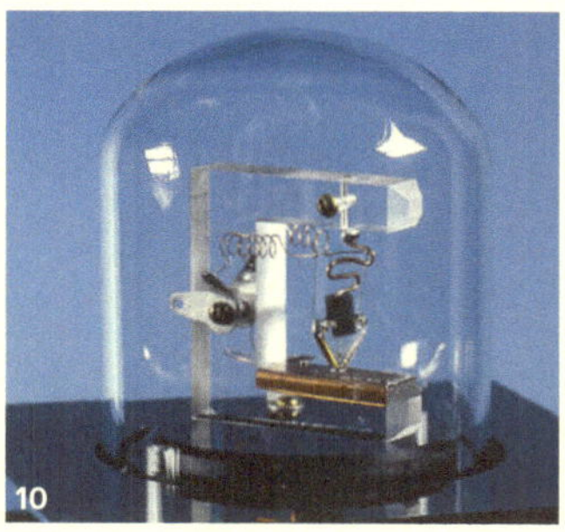
10

1921
— Radio Shack is founded by two brothers, Theodore and Milton Deutschmann, who wanted to provide equipment for the then-nascent field of amateur, or *ham*, radio. 11

1925
— Physicist Julius Edgar Lilienfeld invents a field-effect transistor, a semiconductor used to amplify and switch electronic signals.

1926
— The first transatlantic telephone call is made, from London to New York. 12

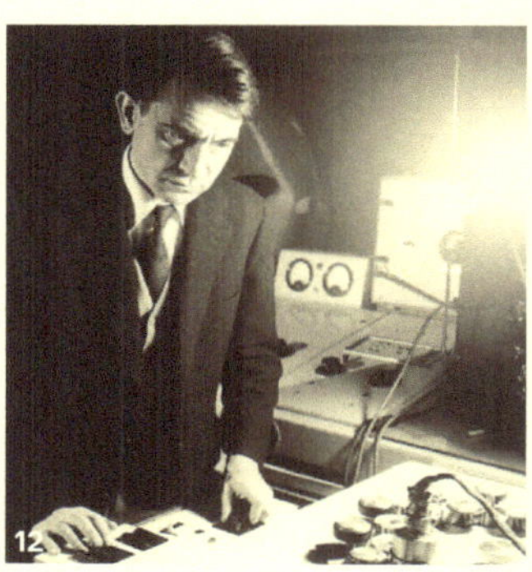
12

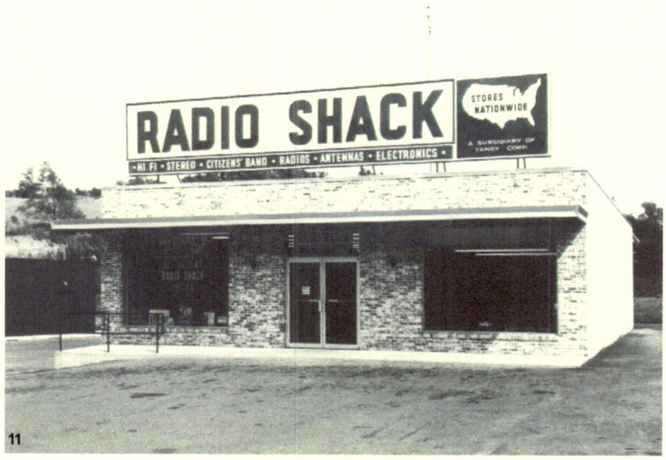

11

1927
— Philo Farnsworth invents the television.

1927
— The first videophone call is produced by AT&T.

1936
— "The Work of Art in the Age of Mechanical Reproduction," which postulates that technology will inevitably alter the way art is perceived, is published by Walter Benjamin.

1937
— Harold "Doc" Edgerton photographs a bullet impacting an apple using stroboscopic equipment.

1939
— The New York World's Fair is the first fair based on the future, with the slogan "Dawn of a New Day." The Fair allows its visitors to take a look at the "World of Tomorrow".

1940
— A ground-based radio navigation system is developed, the predecessor of GPS. 13

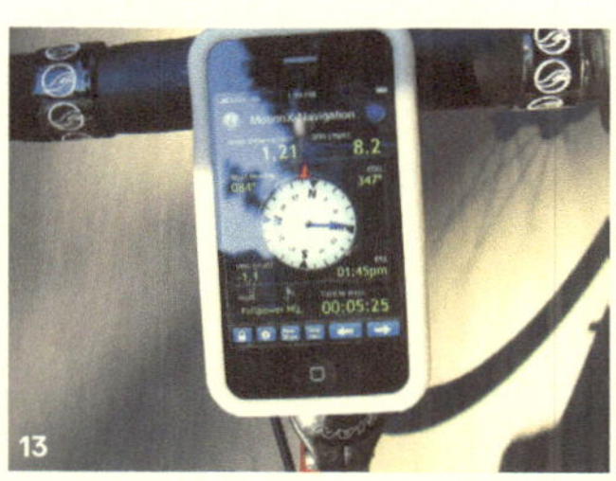
13

1940
— The first solid-body electrical guitar, "the Log," is invented by Les Paul. 14

1946
— The Electronic Numerical Integrator and Computer (ENIAC), the first general purpose electronic computer, is conceived and designed by John Mauchly and J. Presper Eckert of the University of Pennsylvania. It was one thousand times faster than previous electro-mechanical machines, a leap in computing power that no single machine has since matched. 15

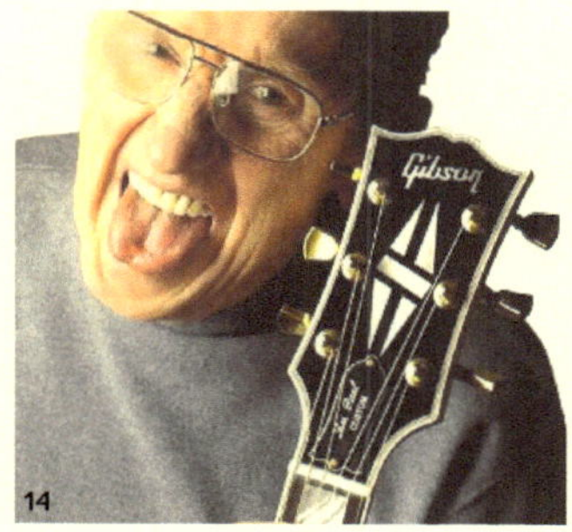

14

1948
— The instant camera is released by the Polaroid Corporation. 16

1948
— Norbert Weiner publishes *Cybernetics*, defining the study of control and communication in the animal and the machine. 17

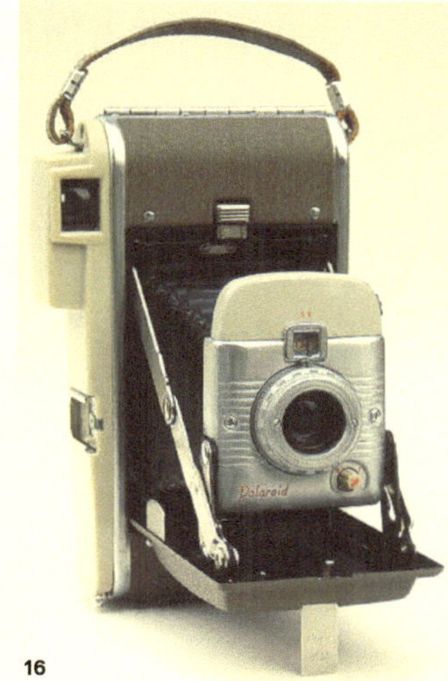
16

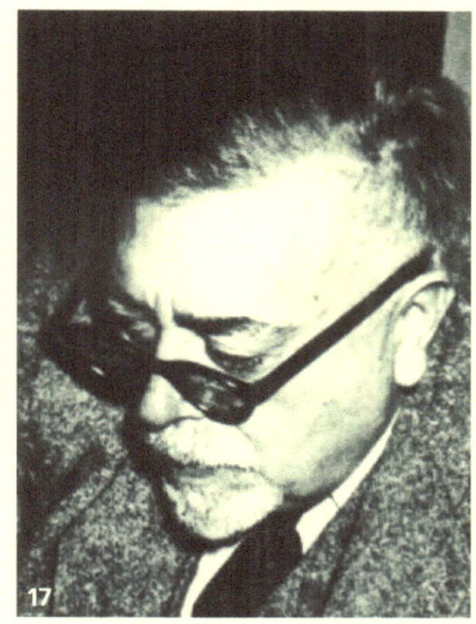
17

1948
— Les Paul's "Lover (When You're Near Me)", his first multi-track recording, is released by Capitol Records.

1949/1986
— The disposable camera is first invented and produced by Photo-Pac as a cardboard camera to be mailed-in for processing. In 1986

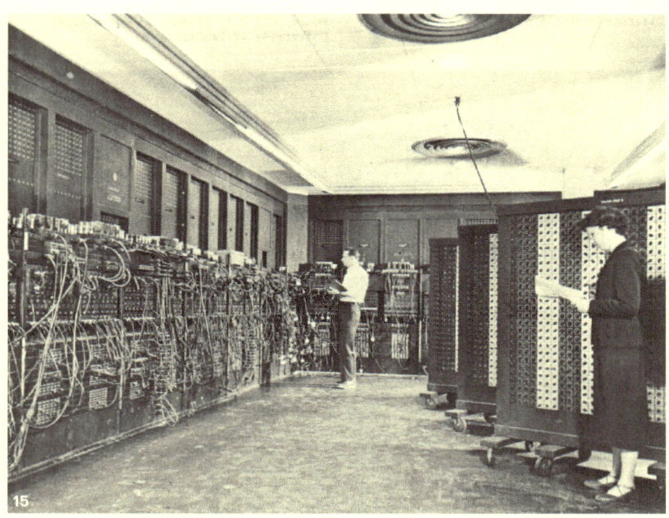
15

and afterwards it is popularized by Fujifilm and Kodak. 18

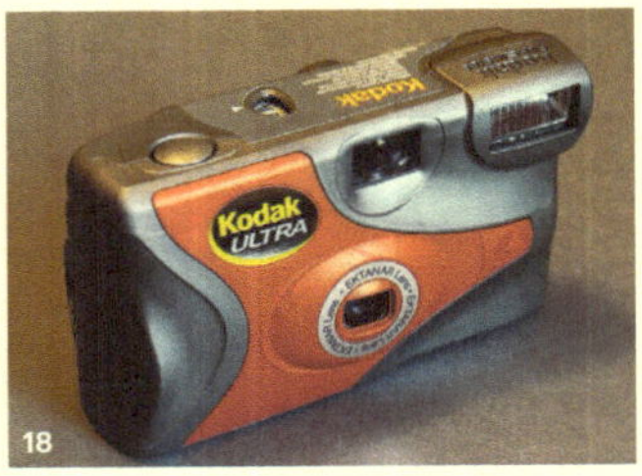

18

c. 1949

— Pierre Schaeffer makes *musique concrète* using magnetic tape at Office de Radiodiffusion-Télévision Française (ORTF).

1950s

1950

— In his paper "Computing Machinery and Intelligence," Alan Turing presents the Turing Test, a means for determining whether a machine is intelligent.

1952

— John Cage releases "Williams Mix," his first magnetic tape work. 19

1953

— The second version of the NTSC (National Television System Committee) broadcast color system is widely adopted.

1953

— Karlheinz Stockhausen's experiments with "Electronic Music" in Pierre Schaeffer's musique concrète studio.

1954

— Buckminster Fuller receives a US patent for the Geodesic Dome, a lattice whose triangular elements distribute stress across the structure.

1954

— Marshall McLuhan publishes *Understanding Media: The Extensions of Man*, a pioneering study in media theory proposing that media themselves, not the content they carry, should be the focus of study—popularly quoted as "the medium is the message".

1956

— Louis and Bebe Barron develop the first entirely electronic score for *Forbidden Planet*, a science fiction film loosely based on William Shakespeare's The *Tempest*.

1956

— The first videotape recorder is developed by Ray Dolby, Charles Ginsberg and Charles Anderson of Ampex.

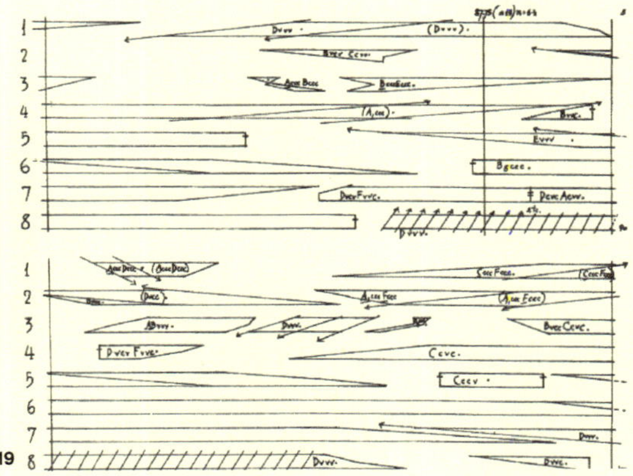

19

1956

— The International Society for the Systems Sciences is established at Stanford.

1957

— Sputnik, the first man-made object to orbit the Earth, is launched; this initiates the Space Age. 20

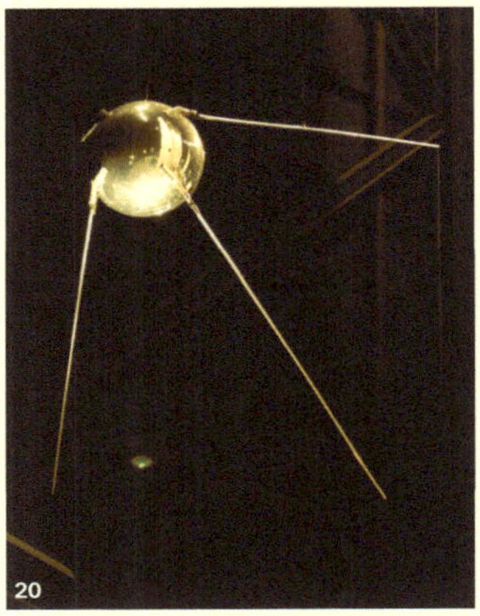

20

1958

— John Whitney, widely considered to be one of the fathers of computer animation, uses customized analog computers to make visual art.

1958

— The National Defense Education Act is signed into law, providing funding to United States education institutions at all levels to increase the number of students, specifically scientists, attending college. The Act was catalyzed by early Soviet success in the Space Race, notably the launch of the first-ever satellite, Sputnik, the year before.

1958

— The *Chicago Tribune* introduced their "Closer Than We Think" strip, reflecting the nation's sentiment shift toward technology-fueled future. 21

1958

— Edgard Varèse releases "Poème électronique," recorded on electronic tape.

1959

— The structure of the DNA molecule is discovered by James D. Watson and Francis Crick.

1959

— The first commercial copier is introduced by Xerox. 22

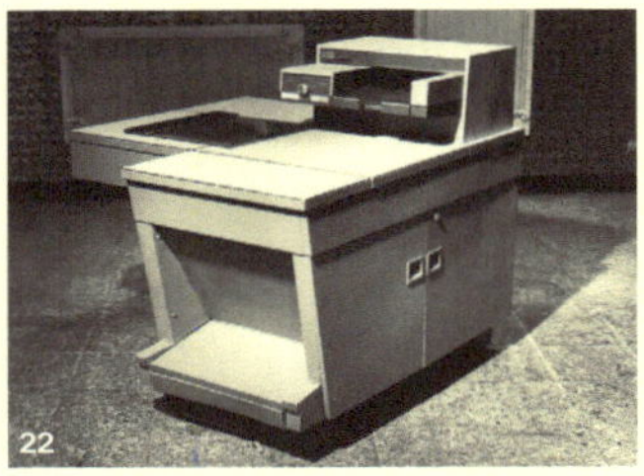

22

1959

— C. P. Snow's lecture, *The Two Cultures*, is delivered at the annual Rede Lecture at the University of Cambridge. The scientist and novelist's thesis focused on the breakdown of communication between the "two cultures" of modern society — the sciences and the humanities — as the major hindrance to solving the world's problems. 23

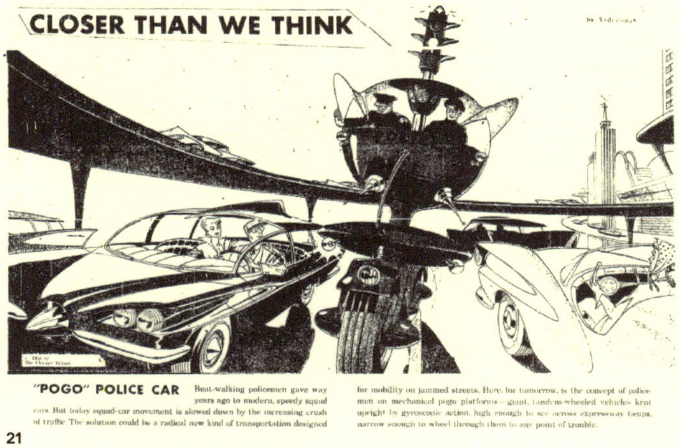

21

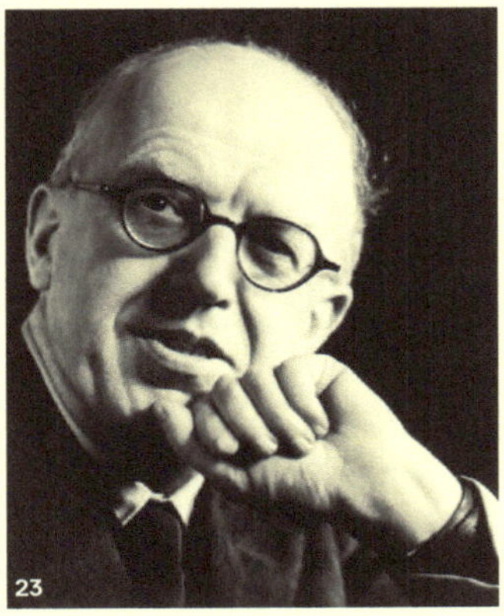

23

1960s

1960-63

— ASCII (American Standard Code for Information Interchange) is developed by The American Standards Association (now the American National Standards Institute-). In 1968, President Lyndon Johnson mandates all government computers support ASCII.

1960

— William Fetter of Boeing coins the term "computer graphics" for his human factors cockpit drawing.

1961

— *Spacewar!*, the first video game, is developed by Steve Russell at MIT for the PDP-1. 24

1962

— The first computer art competition is sponsored by *Computers and Animation*.

1962

— The computer mouse is invented by Douglas Engelbart with the assistance of colleague Bill English. The invention was a small part of Engelbart's larger project of augmenting human intellect.

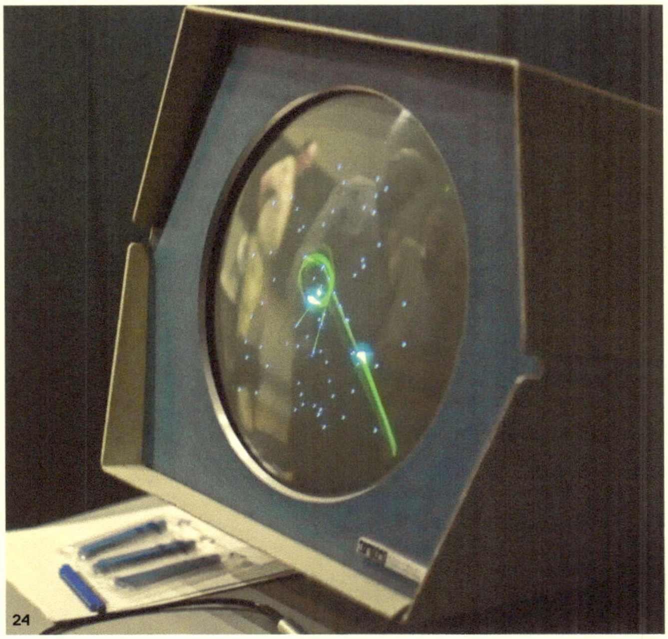

24

1962

— NASA Art Program is established. Artists, including Norman Rockwell and Robert Rauschenberg, are enlisted to present the history of space exploration from their perspective.

1963

— The world's first computer generated film is made by Edward E. Zajac of Bell Laboratories; it demonstrates a gyroscopic system his group had devised to keep an orbiting satellite pointing toward the earth. 25

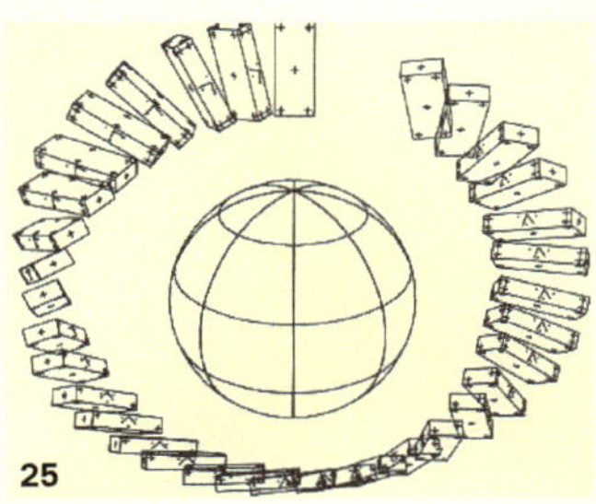

25

1963

— Sketchpad, the first direct graphic interface for drafting, is created by Ivan Sutherland at MIT. 26

1963

— Charles Csuri, a pioneer in the field of computer art, makes his first computer-generated artwork.

1963

— General Dynamics Astronautics publishes *2063 A.D.*, a book containing predictions by scientists, politicians, astronauts and military commanders on the state of space exploration in the year 2063. It is placed in a time capsule. 27

1964

— BASIC, a programming language, is invented by Thomas E. Kurtz and John G. Kemeny at Dartmouth College. 28

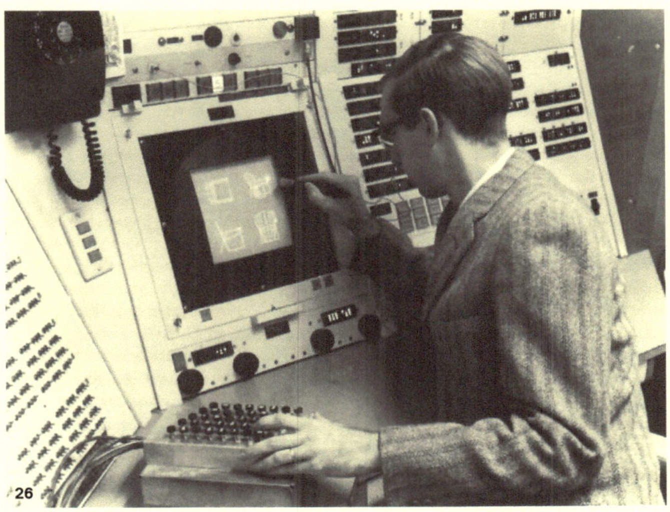

26

1964

— Bell Labs starts an informal artist-in-residence program.

1965

— The first computer art exhibition takes place at Technische Hochschule in Stuttgart.

1965

— The first U.S. computer art exhibition takes place at Howard Wise Gallery in New York.

1966

— Kenneth Knowlton and Leon Harmon, working at Bell Labs, produce one of the first ASCII artworks, "Studies in Perception I".

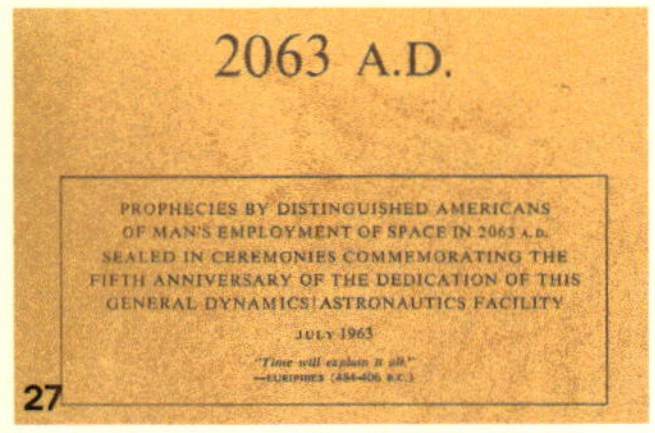

27

1966

— *Nine Evenings of Theater and Engineering*, a series of large-scale performance art collaborations between artists and engineers, takes place in New York at the 69th Regiment Armory building. 29

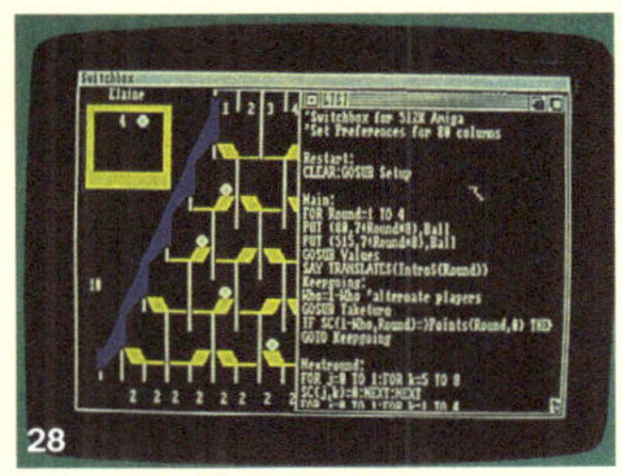

28

1966

— Artists' Placement Group, an organization that actively sought to reposition the role of the artist within a wider social context, including government and commerce, is founded. Its archives are purchased by the Tate Gallery in 2006. 30

1966

— Star Trek, created by Gene Roddenberry, is first broadcast on NBC. 31

1967

— Experiments in Art and Technology (E.A.T.) is founded by Billy Klüver, Robert Rauschenberg, Robert Whitman, and Fred Waldhauer to develop collaborations between artists and engineers.

1967

— The Art and Technology Program at LACMA is founded, pairing contemporary artists with high-tech corporations.

1967

— Artists Robert Irwin and James Turrell experience sensory deprivation in an anechoic chamber at the Jet Propulsion Laboratory.

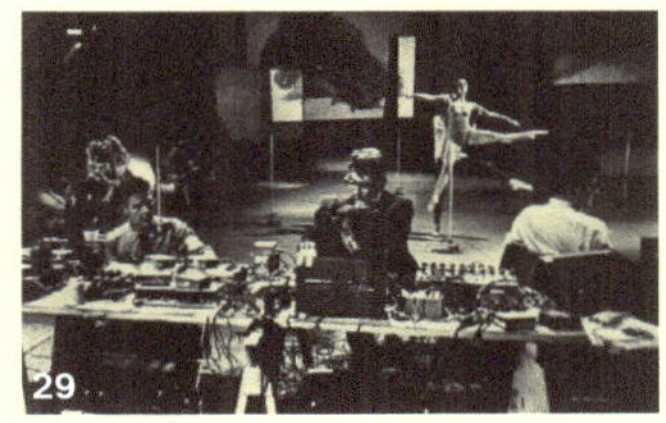

29

1967

— The Center for Advanced Visual Studies at MIT is founded.

1967

— John Whitney becomes artist-in-residence at IBM.

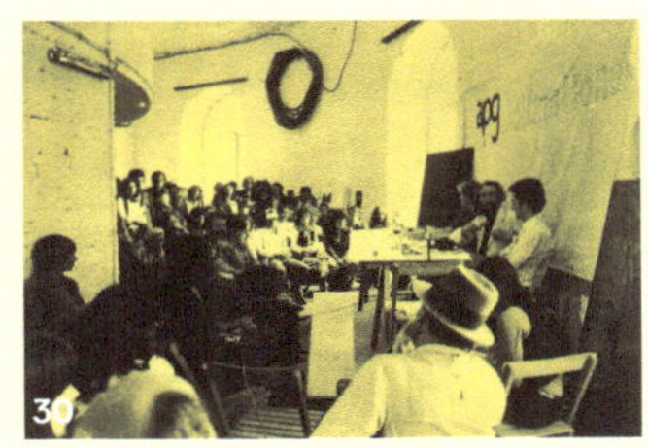

30

1968

— *Cybernetic Serendipity: The Computer and the Arts* exhibition of computer art, curated by Jasia Reichardt, shows at the Institute of Contemporary Arts, London, before touring the U.S. 32

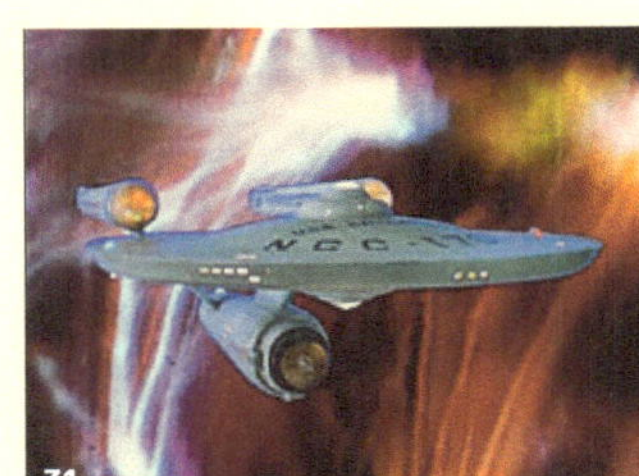

31

1968

— *Leonardo Journal*, for readers interested in the application of contemporary science and technology to the arts, is founded by Frank Malina at MIT. 33

1968
— *2001: A Space Odyssey*, produced and directed by Stanley Kubrick, is released.

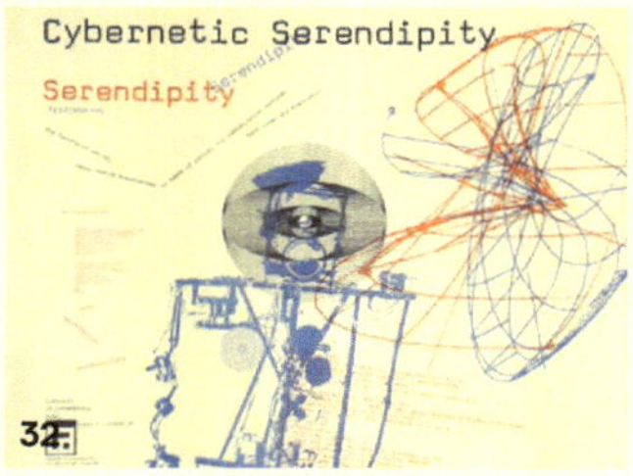

32

1968
— *Whole Earth Catalog*, an American counter-culture catalog, is first published. Published by Stewart Brand, it lists vendors that sold products useful for a creative or self-sustainable lifestyle (e.g. clothing, books, tools, machines, seeds). Steve Jobs has described the catalog as the conceptual forerunner of the World Wide Web, "sort of like Google in paperback form." 34

1968
— The Computer Arts Society is formed as branch of the British Computer Society by John Lansdown and Alan Sutcliffe.

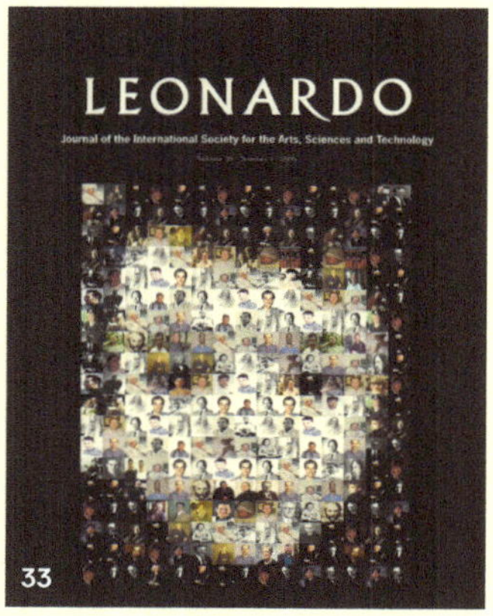

33

1968
— Ant Farm collective is founded by Chip Lord and Doug Michels in San Francisco as an avant-garde architecture, graphic arts, and environmental design practice. The name originates from a friend of the creators who described their work as "underground architecture," which is "what ants do." 35, 36

1969
— The first doctoral dissertation on computer art is submitted at the University of Stuttgart.

1969
— ARPANET, the world's first operational packet switching network, is sold to BBN Technologies.

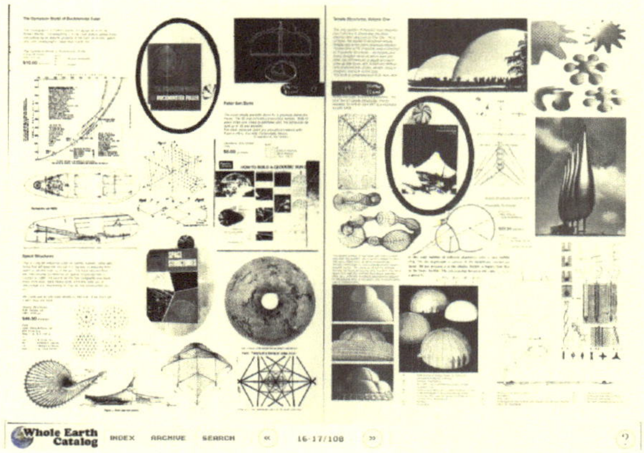

34

1969
— Moon Museum, a miniature ceramic wafer engraved with work by Andy Warhol, Robert Rauschenberg, John Chamberlain, Claes Oldenberg, David Novros and Forrest "Frosty" Myers, is placed on Apollo 12. It is considered the first Space Art object. 37

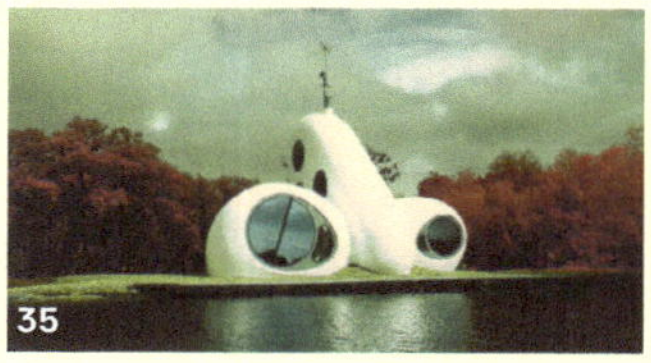
35

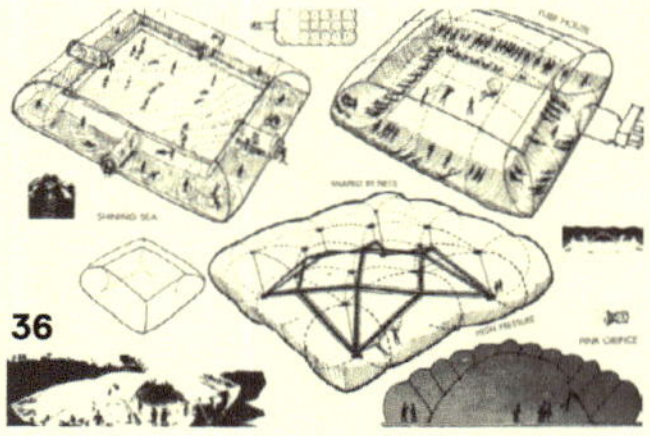
36

1969
— The Exploratorium Museum of science, art and human perception is founded by physicist Frank Oppenheimer in San Francisco, CA. 38

1970s

1970
— The phrase "uncanny valley" is coined by roboticist Masahiro Mori. It states that as a robot's appearance approaches human-like qualities, a human's emotional responses will become positive and empathetic, until a point where the robot is very close to perfect replication. At this point, the response will invert and cause revulsion.

1970
— The historic video magazine *Radical Software* is started by Beryl Korot, Phyllis Gershuny, and Ira Schneider, in the wake of low-cost portable video equipment becoming available to artists and other potential videomakers.

1970
— The first personal computer, the Xerox Alto, is developed by researchers at the Xerox Palo Alto Research Center (PARC). This computer had the first mouse and introduced the now-standard CPU-Monitor- Keyboard-Mouse arrangement.

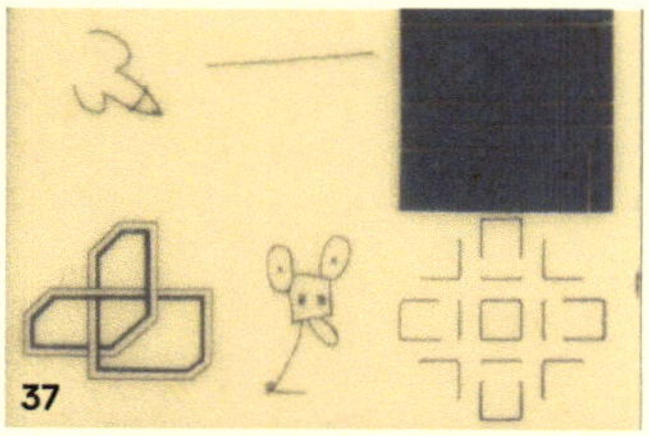
37

1970
— The "Pepsi Cola Pavilion" of the 1970 World's Fair, known as Expo '70, in Osaka, Japan, features work from Experiments in Art and Technology (E.A.T.). 39

1970
— The first Kinko's opens with a single sidewalk copy machine, next to the campus of the University of California, Santa Barbara.

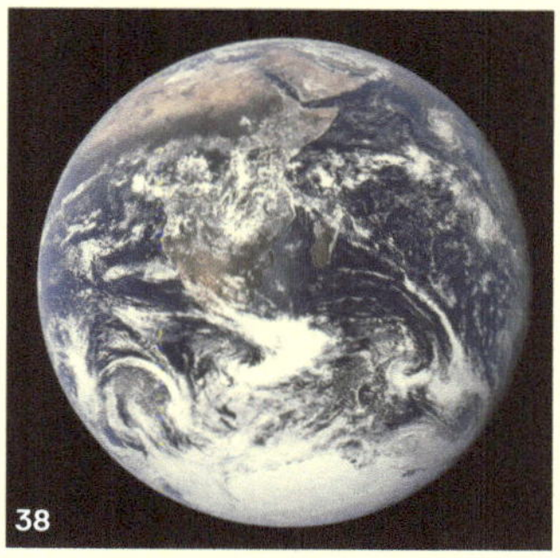
38

1971
— *Computer Space*, the first commercially sold coin-operated video arcade game, is released. Creators Nolan Bushnell and Ted Dabney go on to found Atari the following year.

1971
— A Manfred Mohr exhibition at ARC, the Musée d'Art Moderne de la Ville de Paris, becomes the world's first museum-based solo exhibition of computer generated art.

1971
— Electronic Arts Intermix (EAI), an advocate and resource for video art, is founded in NYC.

1971
— Experimental Television Center, a center dedicated in furthering new work in electronic media technologies, is founded in Owego, NY.

1971
— The Kitchen is founded in NYC by Woody and Steina Vasulka as a collective for video artists, experimental composers, and performers to share ideas with like-minded colleagues.

1972
— The Magnavox Odyssey becomes the first home video game console.

1972-73
— DJ Kool Herc and others establish the nascent genre of hip hop, which is characterized by turntablism, scratching, and rapping.

1974
— The First SIGGRAPH (Special Interest Group on GRAPHics and Interactive Techniques) conference is held.

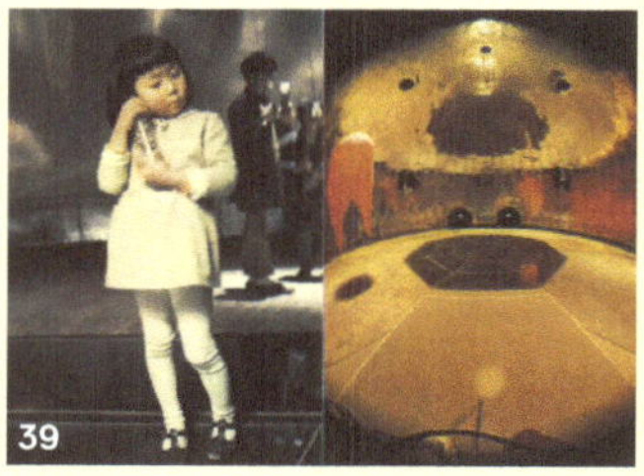
39

1975
— The digital camera is invented by Kodak researcher Steve Sasson. 40

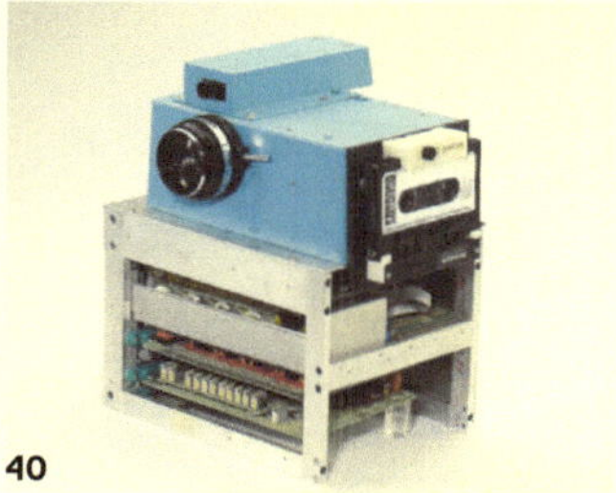
40

1975
— The word "fractal" is coined by mathematician Benoit Mandelbrot. 41

1975
— The first commercial laser printer, Model 3800, is released by IBM. 42

1976
— *Artist and Computer* is published by Ruth Leavitt. 43

1976
— Apple Computer Inc. is founded by Steve Wozniak and Steve Jobs.

1976
— The Exploratorium museum's artist-in-residence program begins with artist Bob Miller's experiments with light and shadow.

1977
— The Apple II, an 8-bit home computer and a major advancement in technology, is released. 44

1977
— *Star Wars*, written and directed by George Lucas, is released. 45

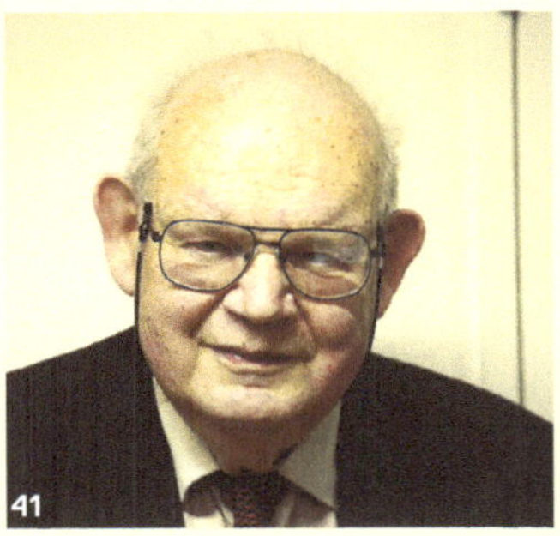
41

1977
— Satellite Arts Project '77, "A Space With No Geographical Boundaries," is conceived by artists Kit Galloway, Sherrie Rabinowitz and collaborators, with the objective of demonstrating that several performing artists separated by geography could appear together in the same live image.

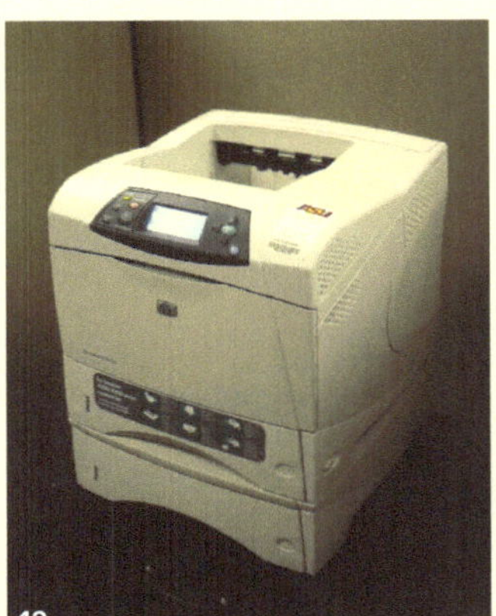
42

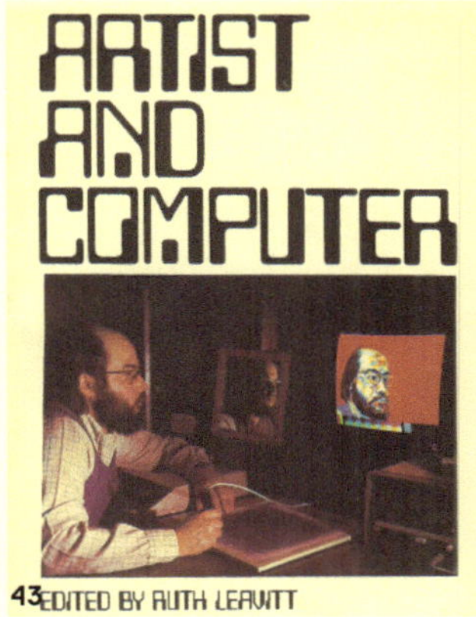

43

1977
— The Voyager Golden Record is launched into space. The record is produced by science writer Timothy Ferris, Carl Sagan, and Anne Druyan; it features music, greetings in 55 languages, and images from Earth. This Earth time capsule is intended for any intelligent extra-terrestrial life form or future humans who may find it. 46

1978
— O'Reilly Media is founded as an American media company that publishes computer books and web sites, and produces conferences on computer technology topics.

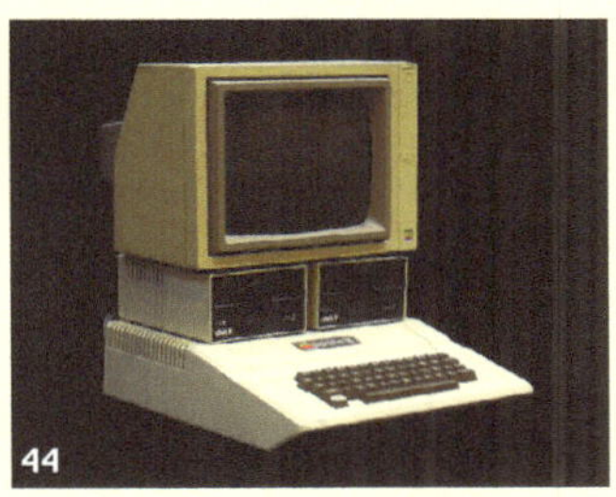
44

1979
— Ars Electronica is created in Linz, Austria, featuring an annual festival for art, technology, and society, as well as a permanent media center and museum.

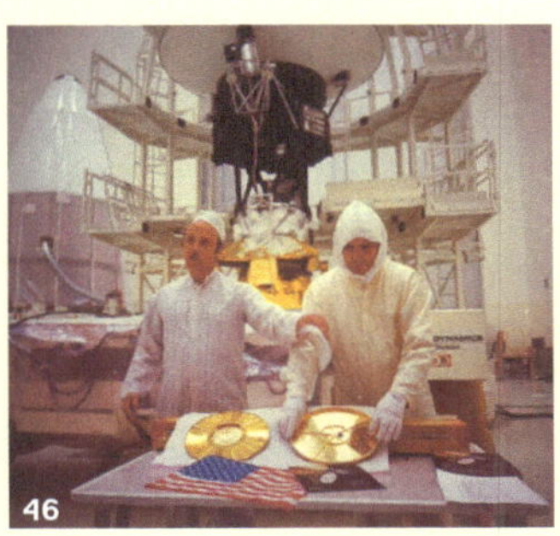
46

1979
— Phillips introduces the first Compact Disc.

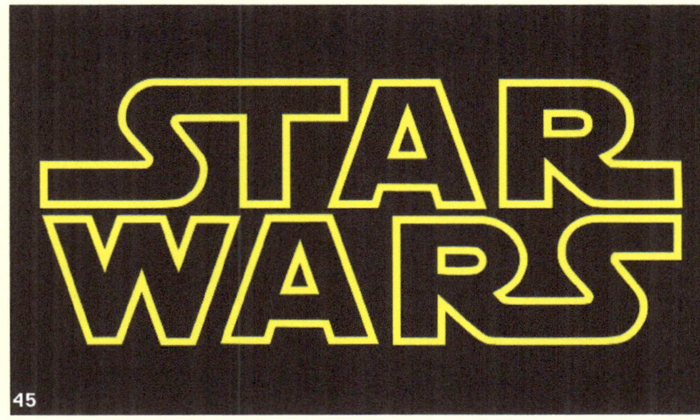

45

1979
— Voyager I photographs Jupiter. 47

1979
— Cable television arrives; Nickelodeon and ESPN are among the first major channels.

1980s

1980
— Quantel introduces Paintbox, a computer graphics workstation used for television graphics.

1980
— Namco releases the arcade game Pac-Man.

1981
— The Space Shuttle Columbia makes its maiden voyage.

1981
— Xerox introduces the Star Computer, launching the concept of Desktop Publishing. 48

1981
— IBM launches its personal computer.

1982
— Disney produces Stephen Lisberger's TRON.

1982
— *Blade Runner*, a film exploring the nature of humanity in a world of artificial intelligence, is released; it's directed by Ridley Scott and based on Philip K. Dick's novel *Do Androids Dream of Electric Sheep?* 49

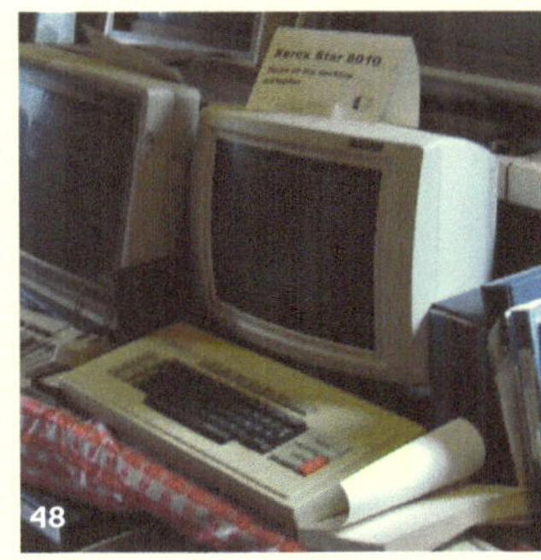
48

1983
— MIDI (Musical Instrument Digital Interface) is invented, enabling electronic musical instruments, computers, and other equipment to communicate and synchronize with each other.

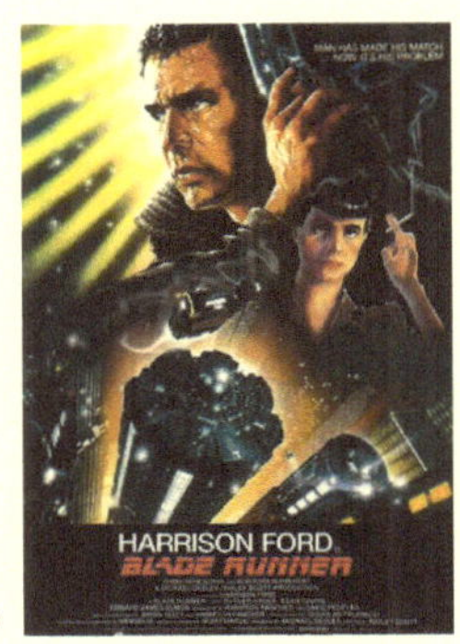

49

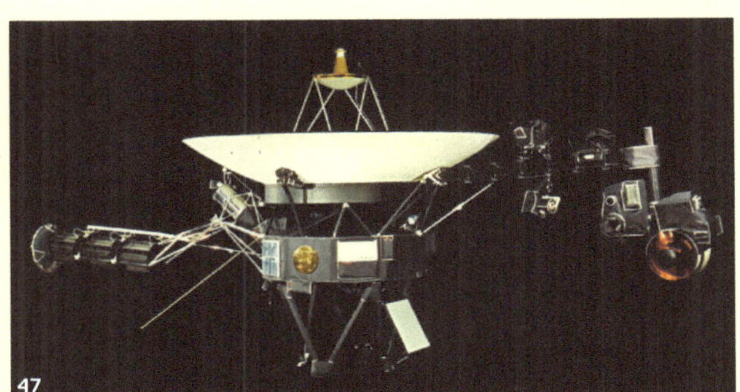
47

1984
— The Graphics Group (a subsidiary of Lucasfilm) releases *The Adventures of Andre and Wally B.*, their first CGI short.

1984
— The Apple Macintosh is first introduced. 50

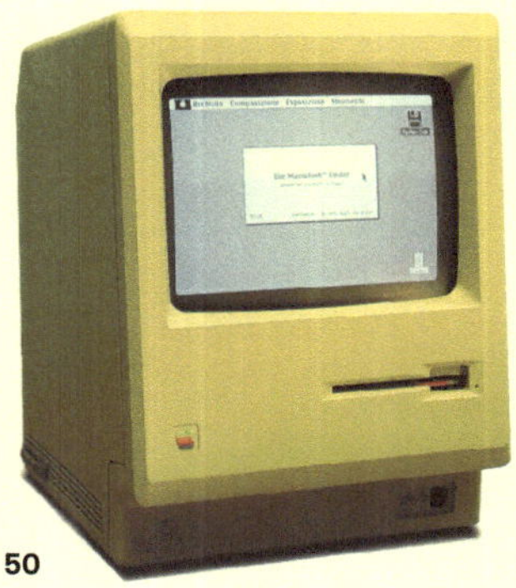

50

1984
— Apple releases MacPaint, a bitmap-based graphics painting program.

1984
— The word "cyberspace" is coined by William Gibson, in his novel, *Neuromancer*.

1985
— The MIT Media Lab is founded by Jerome Weisner and Nicholas Negroponte.

1985
— Lucasfilm's computer division designs the Pixar Computer. Lucas sells the computer division a year later; it then becomes the basis for Pixar Animation Studios.

1987
— Artists Jennifer Bartlett, Richard Hamilton, David Hockney, Howard Hodgkins, Sidney Nolan, and Larry Rivers are invited to use the Qantel Paintbox for BBC 2 TV series *Painting with Light*.

1986
— Andy Warhol uses an Amiga to make portraits of himself and singer Deborah Harry. 51

1986
— *Rendez-vous Houston: A City in Concert* is performed by Jean Michel Jarre using DCA projectors, fireworks, and beamed images to illuminate a kilometer long wall of skyscrapers in downtown Houston.

1986/1997
— Center for Art and Media (Zentrum für Kunst und Medientechnologie, or ZKM) is organized in Karlsruhe, Germany by politicians and representatives of the university, the State Music Academy, the Center for Nuclear Research and other institutions. In 1988, the provincial government votes to establish the Center as a foundation incorporated under public law, and in 1997 it opens its doors.

1987
— The GIF bitmap format is introduced by CompuServe, and the JPEG format is introduced by the Joint Photographic Experts Group.
1989 The world's smallest image, an IBM logo composed of Xenon atoms, is created at IBM; this is the first example of nano-branding.

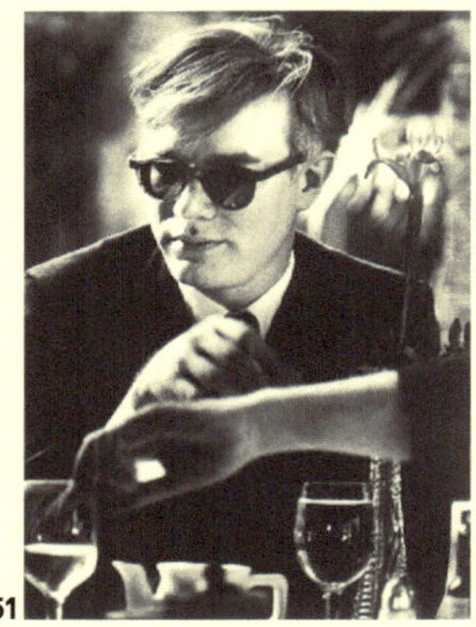

51

1989
— The STUDIO for Creative Inquiry, a center for experimental and interdisciplinary arts, is founded in the College of Fine Arts at Carnegie Mellon University, Pittsburgh, PA.

1990s

1990
—HTML (Hyper Text Markup Language) and the World Wide Web are invented by Tim Berners-Lee. Berners-Lee implements the first successful communication between a Hypertext Transfer Protocol (HTTP) client and server, through the Internet. 52

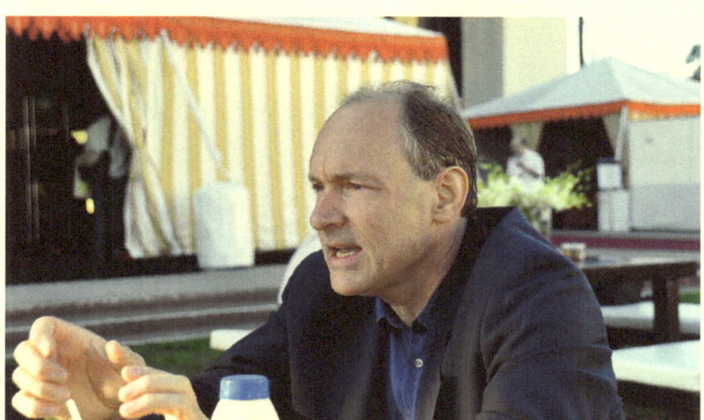

52

1990
— Photoshop 1.0, a computer program that allows users to edit graphics, is developed and published by Adobe Systems Incorporated. 53

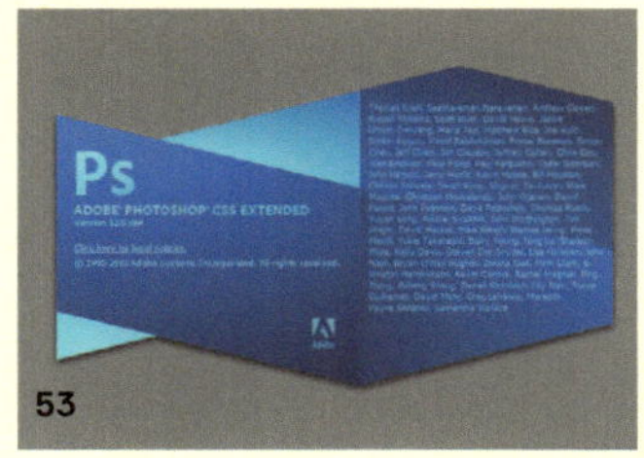

53

1990
— *Batman: Digital Justice*, the first all-digital comic book, is published by DC Comics. 54

54

1990
— ISEA International (formerly Inter-Society for the Electronic Arts) is founded in the Netherlands. It's a nonprofit organization that aims to promote interdisciplinary academic discussion among culturally diverse groups and individuals who work in the areas of art, science, and developing technologies.

1992-2000
— Interval Research Corporation, a laboratory and technology incubator concentrating on consumer product applications and services, with a focus on the Internet, is founded by Paul Allen and David Liddle.

1993
— The Xerox PARC artist-in-residence program is launched.

1993
— The Arts Catalyst (England) is launched, with a goal to commission contemporary art that experimentally and critically engages with science.

1994
— Linux 1.0, a free, open-source operating system, is released. 55

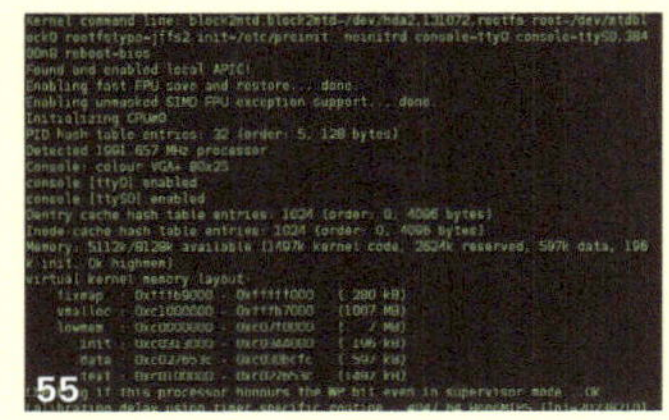

55

1994
— "The World's First Collaborative Sentence," an ongoing project on the World Wide Web by artist Douglas Davis, is commissioned.

1994
— Joachim Sauter creates *Terravision–The Whole Earth Installation*, a virtual representation of the Earth based on satellite images, aerial photos, and altitude/architectural data. Users can navigate from overviews of the Earth to detailed objects in buildings.

1994
— Amazon.com, Inc., a US-based multinational electronic commerce company, is founded.

1995
— The phrase "net.art" is coined by Vuk Cosic, referring to a group of artists that work in the medium of Internet art.

1995
— Jean-Pierre Hébert and Roman Verostko found The Algorists, a formal identity for artists who create "algorithm art," or visual art generated by an algorithm, using a computer. 56

1995
— Internet Phone, the first Voice over IP (VoIP) software, is released by Vocaltec. It utilizes the Internet, as opposed to the public switched telephone network

1995
— *Toy Story*, the first feature-length film made entirely with CGI, is released by Pixar.

1996

— Macromedia Flash 1.0 (now known as Adobe Flash), a multimedia platform used to add animation, video, and interactive features to Web pages, is launched. 57

1996

— Rhizome, a mailing list of new media art, is created. It develops into a not-for-profit arts organization that provides a platform and context for new media work.

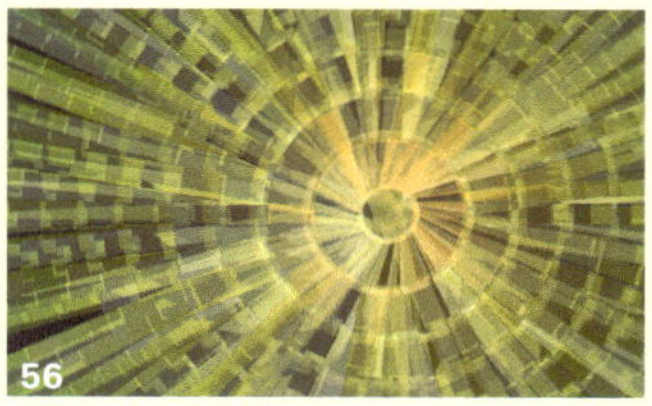
56

1996

— The Long Now Foundation is founded. A private organization that seeks to become a long-term cultural institution providing a counterpoint to what it views as today's "faster/cheaper" mindset, The Long Now Foundation promotes "slower/better" thinking; one of the founding board members is Brian Eno.

1997

— Deep Blue, a chess-playing computer developed by IBM, defeats the reigning World Chess Champion Garry Kasparov.

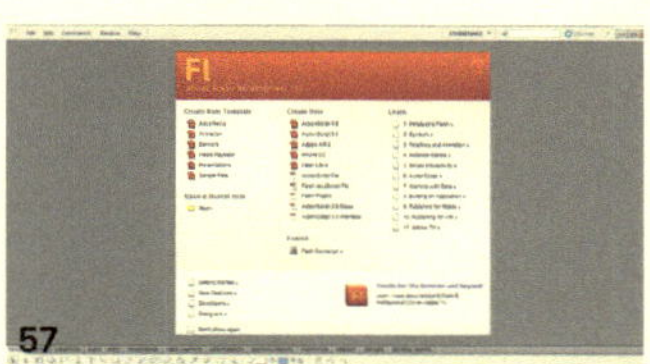
57

1997

— Eyebeam Center for Art & Technology, a nonprofit center located in New York City with the purpose of promoting the creative use of technologies through funding artwork, education, and exhibitions, is created. 58

1997

— Documenta X, the "biggest art show on earth," is directed by Catherine David in Kassel.

1998

— Google, an American multinational public corporation known for its online search services, cloud computing (or Internet-based computing), and advertising applications, is launched. The company hosts and develops several Internet-based services and products, and its main source of profit comes from its advertising program, AdWords.

1998

— Digital Art Museum, an internet resource for the history and practice of digital fine art, is created by Wolfgang Lieser. The website is divided into three sections—Museum, Gallery, and Award—and contains digital art pieces that date back to 1956, as well as artist biographies, articles, and interviews.

2000s

2000

— Dorkbot is founded by Douglas Repetto, as a group of associated organizations that promote grassroots meetings for people involved with electronic art (artists, engineers, designers, scientists, inventors, etc). Their motto is "People doing strange things with electricity."

2000

— Medialab-Prado starts as a small public digital art workshop within the Conde Duque Cultural Center focusing on where art, science, technology, and society intersect.

2000

— Creation of SymbioticA, one of the first research laboratories that enables artists and researchers to engage in wet biology practices in a science department.

2000

— The ZER01 Art and Technology Network is launched in San Jose, CA to encourage creativity at the intersection of art and technology and to produce a festival celebrating this creative intersection.

2001

— Processing (http://www.processing.org), a free and open source "programming language and Integrated Development Environment (IDE) built for the electronic arts and visual design communities," is created by Ben Fry and Casey Reas, meant to teach fundamentals of computer programming within a visual context. As a consequence, the programming learning curve is greatly lessened, providing new opportunities for artists to make work using technology.

2001

— Wikipedia.org, a free online encyclopedia that allows any web user to edit its articles, is launched. It is supported by the non-profit Wikimedia Foundation.

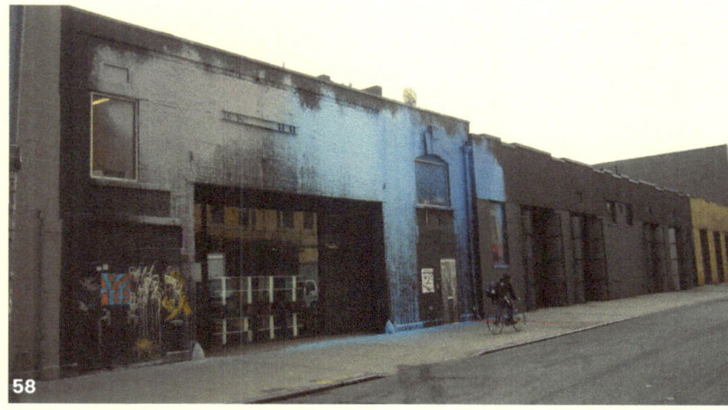
58

2001

— The human genome, the chemical code of chromosomes and genes that make up a human being, is sequenced. 59

2001

— Google Image Search is introduced, allowing users to search the web for image content. Images that have the user's search keyword(s) in their file name, link text, and/or adjacent text make up the search results.

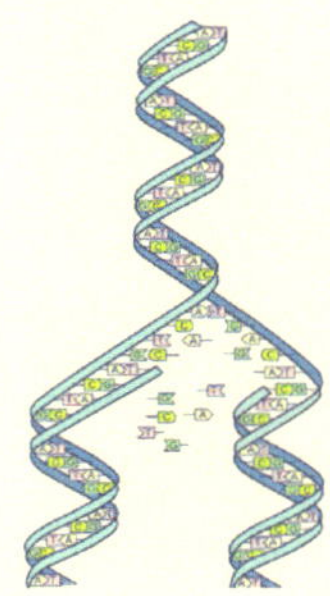
59

2001

— Creative Commons, an American non-profit that works to expand the range of creative works available for others to legally build upon and share, is founded. They release copyright-licenses (Creative Commons licenses) to the public, free of charge. 60

60 creative commons

2001

— C-Level, a cooperative public and private lab formed to share physical, social and technological resources, is founded in Los Angeles.

2002

— The artist Eduardo Kac, with geneticist Louis Houdebine, inserts bioluminescent jellyfish genes into a rabbit, creating glowing "GFP" bunnies, possibly the first transgenic art.

2003

— Second Life, a virtual world developed by Linden Lab, is launched. Users create online identities and interact with one another, free of charge.

2003

— Laurie Anderson, an experimental performance artist and musician, becomes the first artist-in-residence at NASA.

2004

— Machine Project, a Los Angeles-based not-for-profit organization dedicated to making specialized knowledge and technology accessible to artists and the general public, is established.

2004

— The Design Interactions Department at the Royal College of Art opens to students.

2005

— Youtube.com, a website that allows users to upload, view, and share videos, is launched.

61

2005

— *Make* Magazine is founded by O'Reilly Media. It's an American quarterly magazine focused on complex DIY projects involving computers, electronics, robotics, metalworking and woodworking completed with cheap materials and household items.

2005

— Arduino, an open-source electronics prototyping platform based on flexible, easy-to-use hardware and software, is invented by Massimo Banzi and David Cuartielles. 61

2005
— Instructables.com, a community of collaborative users who upload DIY projects that other users can comment on and rate for quality, is created by Eric Wilhelm.

2005
— Google Earth is launched.

2005
— openFrameworks, a coding framework simplified for designers and artists using C++ programming language, is created.

2006
— The first "Maker Faire" is hosted by Make Magazine to celebrate arts, crafts, engineering, science projects and the Do-It-Yourself (DIY) mindset.

2006
— The word "Crowdsourcing" is coined by Jeff Howe in an article for Wired Magazine.

2007
— The iPhone, a smartphone that functions as a camera phone, personal digital assistant, multimedia player, and wireless communication device, is released by Apple Inc.

2007
— Google Lunar X PRIZE, often referred to as GLXP, or Moon 2.0, is launched. It's a competition, sponsored by Google and organized by the X Prize Foundation, which privately-funded spaceflight teams compete to successfully launch, land, and then travel across the Moon's surface with a robot, while sending specified images and data back to Earth. 62

2007
— Artists in Labs (AIL) launches, with the aim of promoting knowledge transfer between artists and scientists. The AIL program is a Swiss partnership between the Zurich University of the Arts, the Institute of Cultural Studies (ICS) and the Bundesamt für Kultur BAK.

2008
— The Science Gallery, a public science centre at Trinity College, Dublin, Ireland, opens, presenting various exhibitions and lectures with a view to science outreach.

2008
— *The Age of Wonder: How the Romantic Generation Discovered the Beauty and Terror of Science*, a book inspired by the scientific ferment that swept through Britain at the end of the 18th century, is written by Richard Holmes. The author charts the many voyages of discoveries–astronomical, chemical, poetical, philosophical–and describes how this period formed the basis for modern scientific discoveries.

62

2009
— MakerBot Industries is founded. The company sells Cupcake CNC machines, an open source, 3D printer rapid prototyping machine that makes affordable desktop 3D printing accessible at home, as do it yourself kits.

2009
— Kickstarter is founded by Perry Chen, Yancey Strickler, and Charles Adler, as a "crowd-funding" site that helps launch creative projects that meet a threshold pledge goal.

2010s

2010
— The iPad is released by Apple, Inc. The tablet computer's primary function is to be a platform for audio-visual media, such as books, periodicals, music, movies, games, and web-based content.

2010
— Kinect is released by Microsoft for the Xbox 360 video game platform; it allows users to control and interact with an Xbox 360 without the use of controllers, but through gestures and spoken commands. Kinect's goal is to broaden the Xbox's user audience beyond the average video gamer. Hackers and artists immediately repurpose the technology.

2010
— The National Fab Lab Network Act is proposed by Representative Bill Foster to provide for the establishment of a network of fabrication labs across the United States (one for every 700,000 people or about 443 total) to foster a new generation with scientific and engineering skills.

2011
— *The Art as a Way of Knowing Conference*, organized by the Exploratorium, sponsored by a grant from the National Science Foundation, occurs. Artists, scientists, curators, writers, and educators from around the world discuss the role of arts in public interdisciplinary learning environments.

2011
— *New Art/Science Affinities* book sprint takes place at Carnegie Mellon University.

BIBLIOGRAPHY & RESOURCES

Aceti, Lanfranco “About Leonardo/ISAST.” http://www.leonardo.info/isast/isastinfo.html

Belanger, Matthew. “Network Culture and the Poetics of Reality.” http://varnelis.networked-book.org/The-immediated-now-network-culture-and-the-poetics-of-reality

Bell, Sherry and Langdon Morris, eds. *Living in Space: Cultural and Social Dynamics, Opportunities, and Challenges and Permanent Space Habitats*. An Aerospace Tehnology Working Group Book, 2009.

Blais, Joline and Jon Ippolito. *At the Edge of Art*. Thames & Hudson, 2006.

Blumenthal, Marjory S, Alan S. Inouye, and William J. Mitchell, eds. *Beyond Productivity: Information Technology, Innovation, and Creativity*. The National Academies Press, 2003.

“Bibliography.” http://tate.org.uk/learning/artistsinfocus/apg/bibliography.htm
“Brave New Worlds”. Walker Art Center. 1750 Hennepin Ave, Minneapolis. October 2007.

Bunch, Bryan and Alexander Hellemans, eds. *The Timetables of Technology: A Chronology of the Most Important People and Events in the History of Technology*. Simon & Schuster, 1994.

Buxton, Bill. “The Role of the Artist in the Lab.” Meisterwerke der Computer Kunst, June 1988. http://billbuxton.com/artistRole.html

Cook, Sarah, Gfader, Verina, Graham, Beryl and Lapp, Axel, *A Brief History of Working with New Media Art - Conversations with Artists*, Berlin: The Green Box, 2010.

Crawford, Michael. *Shop Class as Soulcraft: An Inquiry Into the Value of Work*. Penguin Press HC, 2009.

Daniel, Dieter and Barbara Schmidt, eds. *Artists as Inventors – Inventors as Artists*. Hatje Cantz, 2008.

Davis, Douglas. *Art and the Future: A History / Prophecy of the Collaboration between Science, Technology and Art*. Praeger Publishers, 1973.

Dessau, Weimer and Hans M. Wingler. *Bauhaus*. MIT Press, 1969.

Donald, Diana, and Jane Munro, ed. *Endless Forms: Charles Darwin, Natural Science and the Visual Arts*. Yale University Press, 2009.

Evers, Frans, Lucas van der Velden and Jane Peter van der Wenden, eds. *The Art of Program: Sonic Acts 2001 Conference on digital art, music and education*. 2001.

Fifield, George. “A Mutual Attraction: Creativity and Innovation.” www.bostoncyberarts.com/festival/2007/home_rr_air.php

Fox, Howard N. “In Context: LACMA's Art and Technology Program, 1967-1971.” Los Angeles Country Museum of Art. http://collectionson-line.lacma.org/mwebcgi/mweb.exe?request=epage;id=502074;type=803

Frauenfelder, Mark, ed. *MAKE: Technology on your time*. Volume 22.

Genter, Robert. *Late Modernism: Art, Culture, and Politics in Cold War America*. University of Pennsylvania Press, 2010.

Glimcher, Marc. *Logical Conclusions: 40 Years of Rule-Based Art*. Pacewildenstein, 2005.

Goldring, Elizabeth and Otto Piene, eds. *Centerbeam*. MIT, 1980.

Gordinier, Jeff. “Insect Dinners- Waiter, There's Soup in my Bug.” http://nytimes.com/2010/09/22/dining/22bug.html

Gough, Maria. *The Artist as Producer*. University of California Press, 2005.

Harris, Craig, ed. *Art and Innovation: The Xerox Parc Artist – In – Residence Program*. MIT Press, 1995.

Jana, Reena and Mark Tribe. *New Media Art*. Taschen, 2006.

Joffe, Josef. “America's Secret Weapon: First Chapter: ‘The Cultural Cold War’.” http://nytimes.com/books/00/04/23/reviews/000423.23joffet.html?_r=1&sq=CIA%20Abstract%20Expresionism&st=cse

Kepes, Gyorgy. *Language of Vision*. Paul Theobald and Company, 1944.

Kepes, Gyorgy. *The New Landscape in Art and Science*. Cambridge, 1956.

Kepes, Gyorgy, ed. *Arts of the Environment*. George Braziller, 1972.

Kepes, Gyorgy, ed. *Education of Vision*. George Braziller, 1965.

Kepes, Gyorgy, ed. *Structure in Art and Science*. George Braziller, 1965.

Kepes, Gyorgy, ed. *The Man-made Object*. George Braziller, 1966.

Kepes, Gyorgy, ed. *The Nature of Art of Motion*. George Braziller Inc, 1965.

Klein, Julie Thompson. *Interdisciplinary: History, Theory, and Practice*. Wayne State University Press, 1990.

Klüver, Billy, Julie Martin, and Barbara Rose, ed. *Pavilion by Experiments in Art and Technology*. Experiments in Art and Technology, 1972.

Klüver, Billy and Robert Rauschenberg. “Brief Description of E.A.T.” E.A.T. news Volume 1, NO.2. 1 June 1967.

Kranz, Stewart, comp. and Margaret Holton, ed. *Science and Technology in the Arts: A Tour through the realm of Science/Art*. Van Nostrand Reinhold Company, 1974.

Lambert, Nick. “Early Computer Art in Context: Computer Art Thesis.” http://thesis.lamberts-blog.co.uk/?page_id=128

Leavitt, Auth, ed. *Artist and Computer*. Harmony Books, 1976.

Lee, Pamela M. *Chronophobia on Time in the Art of the 1960s*. MIT, 2004.

Leonardo/The International Society for the Arts, Sciences and Technology, http://www.leonardo.info/

Lieser, Wolf. *The World of Digital Art*. H. F.mUlmann, 2010.

Media, New Media, Postmedia, Domenico Quaranta, Postmediabooks, Milan 2010 http://medianewmediapostmedia.wordpress.com/english-abstract/

Moran, Joe. *Interdisciplinary: The New Critical Idiom*. Routledge, 2002.

Morris, Catherine, curator. *9 Evenings Reconsidered: art, theatre, and engineering, 1966*. MIT List Visual Arts Center, 2006.

Nawi, Diana. “Margorie Susman Curatorial Fellow at the Museum of Contemporary Art, Chicago.” http://studiochicago.blogspot.com/2010/03/factory-as-studio-as-art.html 15 March 2010.

Nelson, Kevin. “In Response to ‘Innovation, Education and Makers’.” http://soundcyst.com/blog/2010/10/18/in-response-to-innovation-education-and-makers/

NTT InterCommunication. *E.A.T.- The Story of Experiments in Art and Technology*.

Reichardt, Jasia, ed. *Cybernetic Serendipity*. Frederick A. Praeger, 1969.

Oppenheimer, J. Robert. “Words That Men Live By.” Federal Observer 11.49 (2011). http://www.federalobserver.com/words.php?words=11850

"Overview." Tate, APG: Artists Placement Group – Overview. http://tate.org.uk/learning/artsinfocus/apg/overview.htm

Palais de Tokyo, *What do you Expect from an Art Institution in the 21st Century*. Palais de Tokyo, 2001.

Pescovitz, David. "White House's Tom Kalil on Maker Culture." Boing Boing. 5 October 2010. http://m.boingboing.net/2010/10/05/white-houses-tom-kal.html

Radar, O'Reilly. "Innovation, Education and Makers." http://radar.oreilly.com/2010/10/innovation-education-and-the-m.html

"Rate of Return: The Artist Placement Group: View of 'Inno7o' 1971." http://artforum.com/inprint/issue=201008&id=26419

Reichardt, Jasai. *The Computer in Art*. Studio Vista/Van Nostrand Reinchold Paperback, 1971.

Reichle, Ingeborg. *Art in the Age of Technoscience: Genetic Engineering, Robotics, and Artificial Life in Contemporary Art*. Springer Wien New York, 2009.

Saunders, Francis Stonor. "Modern Art was CIA 'Weapon' – World News." http://independent.co.uk/news/world/Modern-art-was-cia-weapon-1578808.html

Schopf, Christine and Herausgegeben von Gerfried Stocker, eds. *A New Culture Economy: the Limits of Intellectual Property*. Ars Electronica, 2008.

Schopf, Christine and Herausgegeben von Gerfried Stocker, eds. *Human Nature: Arts Electronica*. Ars Electronica, 2009.

Shanken, Edward A. "Artists in Industry and the Academy: Collaborative Research, Interdisciplinary Scholarship and the Creation and Interpretation of Hybrid Forms."

LEONARDO. Vol. 38, No. 5, 2005. Pp. 415-418.

Snow, C.P. *The Two Cultures: A Second Look*. Cambridge, 1965.Strosberg, Eliane. Art and Science. Abbeville Press Publishers, 1999.

Sterling, Bruce, eds. *Mirrorshades: The Cyberpunk Anthology*. Arbor House, 1986.

Talley-Jones, Kathy. *Overview of the Program*. Los Angeles County Museum of Art, 2008.

Tribe, Mark and Jana, Reena. *New Media Art*, Taschen, 2006.

Tuchman, Maurice. *A Report on the Art and Technology Program of the Los Angeles County Museum of Art 1967-1971*. Los Angeles County Museum of Art, 1971.

UCLA. *The Artist as Inventor*. Center for Medieval and Renaissance Studies.

Vasulka, Woody. *The Brotherhood*. NTT InterCommunication Center, 1998.

Wands, Bruce. *Art of the Digital Age*. Thames & Hudson, 2006.

Watts, Patricia. "Make:Craft." http://themakershow.blogspot.com

Weschler, Lawrence. *Seeing is Forgetting: The Name of the Thing One Sees: A Life of Contemporary Artist*. University of California Press, 1982.

Wikipedia—The Free Encyclopedia. "John Dewey." http://en.wikipedia.org/wiki/John_Dewey

Wilson, Stephen. *Art and Science: How scientific research and technological innovation are becoming key to 21st century aesthetics*. Thames & Hudson, 2010.

Wilson, Stephen. *Information Arts: Intersections of Art, Science and Technology*. MIT Press, 2002.

RESOURCES

Artsactive
www.artsactive.net/en

CRUMB, Curatorial Resource for Upstart Media Bliss
www.crumbweb.org

Digicult
www.digicult.it/en

Digitalarti
www.digitalarti.com

Hackerspaces
www.hackerspaces.org/wiki/Hackerspaces

Instructables
www.instructables.com

Journal for Artistic Research
www.jar-online.net

Leonardo Online
www.leonardo.info

Make Magazine
www.makezine.com

Networked Performance
www.turbulence.org

Neural.it
www.neural.it

N E W M E D I A A R T . E U
www.newmediaart.eu

Random
www.random-magazine.net

RealTime Arts
www.realtimearts.net

Rhizome
www.rhizome.org

we-make-money-not-art
www.we-make-money-not-art.com

we-need-money-not-art
www.we-need-money-not-art.com

EVENTS AND FESTIVALS

Alpaaville, London
www.alphavillefestival.co.uk

AND - Abandon Normal Devices, Across Northwest England
www.andfestival.org.uk

Ars Electronica, Linz
www.aec.at

Artefact Festival, Leuven
www.artefact-festival.be

Art Futura, Barcelona - Buenos Aires
www.artfutura.org/v2

AV Festival, East Midlands, UK
www.avfestival.co.uk

Boston Cyberarts Festival
www.bostoncyberarts.org/festival

Chaos Communication Congress, Berlin
www.events.ccc.de/congress/2010/wiki/Welcome

Cimatics, Brussels
www.cimatics.com

Conflux festival, New York
www.confluxfestival.org

Digital Art Festival, Taipei
www.dac.tw/daf10/home_en.html

Dorkbot, worldwide
www.dorkbot.org

Electrofringe, Newcastle,
New South Wales, Australia
www.electrofringe.net

Electrohype, Malmö
www.electrohype.org

Elektra Festival, Montreal
www.elektramontreal.ca

Enter Festival, Prague
www.festival-enter.cz

European Media Art Festival, Osnabrück
www.emaf.de

Experimenta, International Biennial of Media Art, across Australia
www.experimenta.org

Eyeo, Minneapolis
www.eyeofestival.com

FILE - Electronic Language International Festival
www.file.org.br

The Finnish Bioart Society
www.bioartsociety.fi

FutureEverything, Manchester
www.futureeverything.org

The Influencers, Barcelona
www.theinfluencers.org

Interferenze, Bisaccia
www.interferenze.org

ISEA - Inter-Society for the Electronic Arts.
www.isea-web.org

Japan Media Arts festival, Tokyo
www.plaza.bunka.go.jp/english

Mal au Pixel, Paris
www.craslab.org/malaupixel/2010

Maker Faire
www.makerfaire.com

OFFF, Barcelona
www.offf.ws/2

Piemonte Share, Turin
www.toshare.it

Piksel, Bergen
www.piksel.no

Pixelache, Helsinki
www.pixelache.ac

Shift Festival, Basel
www.shiftfestival.ch

SONAR, International Festival of Advanced Music and Multimedia Art, Barcelona
www.2011.sonar.es

Spark, Minneapolis
www.spark.umn.edu

STRP, Eindhoven
www.strp.nl

Subtle Technologies, Toronto
www.subtletechnologies.com

The Next HOPE conference, New York
www.thenexthope.org

TodaysArt, The Hague
portal.todaysart.nl

Transitio, Mexico DF
transitiomx.net

Transmediale, Berlin
www.transmediale.de

WRO International Media Art Biennale, Wrocław
www.wrocenter.pl

ZEMOS98 Festival, Sevilla
www.13festival.zemos98.org

01SJ Biennial, San Jose
www.01sj.org

INSTITUTIONS, ORGANIZATIONS AND EXHIBITION SPACES

Ars Electronica Center, Linz
www.new.aec.at/news

Artists in Labs, Zurich
www.artistsinlabs.ch

The Arts & Genomics Centre, Leiden
www.artsgenomics.org

Atelier Nord, Oslo
www.ateliernord.no

ANAT (Australian Network for Art and Technology), Adelaide
www.anat.org.au

ATNE, Art and Technology New England, Boston
www.atne.org

AXIOM Center for New and Experimental Media, Jamaica Plain
www.axiomart.org

The Banff Centre, Alberta
www.banffcentre.ca

UCI Beall Center for Art and Technology
www.beallcenter.uci.edu
BEK, Bergen Center for Electronic Arts
www.bek.no

C3. Budapest
www.c3.hu

Capsula [art, science, mature],
Helsinki/Barcelona
www.00capsula00.wordpress.com

Centro Multimedia", center for research on art and technology, Mexico DF
www.cmm.cenart.gob.mx

Daniel Langlois Foundation for Art, Science, and Technology, Montreal
www.fondation-langlois.org

Edith Russ Haus, Oldenburg
www.edith-russ-haus.de

EMPAC, The Curtis R. Priem Experimental Media and Performing Arts Center, Troy
www.empac.rpi.edu

Exploratorium, San Francisco
www.exploratorium.edu

Eyebeam Art and Technology Center, New York
www.eyebeam.org

FACT, Liverpool
www.fact.co.uk

GAFFTA, Gray Area Foundation For The Arts, San Francisco
www.gaffta.org

HKMV - Hartware MedienKunstVerein, Dortmund
www.hmkv.de

HTTP gallery, London
www.http.uk.net

ICC (NTT InterCommunication Center), Tokyo
www.ntticc.or.jp

iMAL, Brussels
www.imal.org

InterAccess Electronic Media Arts Centre, Toronto
www.interaccess.org

The Institute of Unnecessary Research, Brighton
www.unnecessaryresearch.org

The Israeli Center of Digital Art, Holon
www.digitalartlab.org.il

Kitchen Budapest, Budapest
www.kitchenbudapest.hu/en

Laboral, Center for Art and Industrial Creation, Gijón
www.laboralcentrodearte.org

Le Laboratoire, Paris
www.lelaboratoire.org

Laboratorio Arte Alameda, Mexico DF
www.artealameda.bellasartes.gob.mx

Laboratory Art Beijing
www.bjartlab.com

Ljudmila - Ljubljana Digital Media Lab, Ljubljana
www.ljudmila.org

Machine Project, Los Angeles
machineproject.com

Medialab-Prado, Madrid
www.medialab-prado.es

Mediamatic, Amsterdam
www.mediamatic.net

Nabi Art Center, Seoul, Korea
www.nabi.or.kr

New Media Caucus, worldwide
www.newmediacaucus.org

New Media Center_kuda.org, Novi Sad, Serbia
www.kuda.org

New Media Scotland, Edinburgh
www.mediascot.org

NIMk, Nederlands Instituut voor Mediakunst, Amsterdam
www.nimk.nl

OBORO, Montreal
www.oboro.net

[plug.in], Basel
www.iplugin.org

RIXC - The Centre for New Media Culture, Riga
www.rixc.lv

SAT, Society for Arts and Technology, Montreal
www.sat.qc.ca

Science Gallery, Dublin
www.sciencegallery.com

SAT, Society for Arts and Technology, Montreal
www.sat.qc.ca

SymbioticA, Perth
www.symbiotica.uwa.edu.au

STUDIO for Creative Inquiry, Carnegie Mellon University, Pittsburgh
www.studioforcreativeinquiry.org

The Arts Catalyst, London
www.artscatalyst.org

The BioArt Initiative at Rensselaer, Troy
www.arts.rpi.edu/bioart

The LAB, San Francisco
www.thelab.org

The Leonardo, Salt Lake City
www.theleonardo.org

Upgrade, worldwide
www.theupgrade.net

Vooruit, Ghent
www.vooruit.be

V2_Institute for the Unstable Media, Rotterdam
www.v2.nl

Waag Society, Amsterdam
www.waag.org

ZKM, Karlsruhe
www.on1.zkm.de

AWARDS

Buckminster Fuller Challenge
www.challenge.bfi.org

FutureEverything Award
www.award.futureeverything.org

Prix Ars Electronica
www.new.aec.at/prix/en/about

VIDA, Art and Artificial Life International Awards
www.fundacion.telefonica.com/es/arteytecnologia/certamen_vida

EDUCATION

Academy of Media Arts Cologne
www.khm.de

Art and Technology Program, Ohio State, Columbus
www.artandtech.osu.edu

Berkeley Center for New Media
www.cnm.berkeley.edu

Carnegie Mellon University BXA Intercollege Degree Program, Pittsburgh
www.cmu.edu/interdisciplinary/programs/index.html

Design Interactions at the RCA, London
www.interaction.rca.ac.uk

Digitale Klasse (Digital Media Design at the University of Arts), UdK, Berlin
www.digital.udk-berlin.de

Ecole Supérieure d'Art d'Aix-en-Provence
www.ecole-art-aix.fr

Grupo de investigación en bioarte de la escuela de artes plásticas de la universidad nacional de Colombia, Bogota
www.comohacerunafresadeltamanodeunasandia.wordpress.com

Hexagram: Institute for Creation/Research in Media Arts and Technologies, Montreal
www.finearts.concordia.ca/research/fineartsresearch/researchcentres/#hexagram

iEAR (Integrated Electronic Arts at Rensselaer), Troy
www.arts.rpi.edu

IAMAS (International Academy of Media Arts and Sciences), Ogaki City, Gifu
www.iamas.ac.jp

ITP (Interactive Telecommunications Program) Tisch School of the Arts, New York
www.itp.nyu.edu

Srishti School of Art, Design and Technology, Bangalore
www.srishti.ac.in

University of California, Los Angeles, Design Media Arts, Los Angeles
www.dma.ucla.edu

University of California, Irvine, ACE (Arts, Computation and Engineering)
www.ace.uci.edu

IMAGE SOURCES

All reasonable attempts to properly credit artists and their work have been made. Corrections will be made in future additions.

EXTREME WRITING
1. Courtesy Jonathan Minard and STUDIO for Creative Inquiry

STATEMENT, EXPERIMENTS IN ART AND TECHNOLOGY
2. Statement of purpose, Experiments in Art and Technology, (New York, United States), 1967. E.A.T. aims drafted by Billy Klüver and Robert Rauschenberg. via Daniel Langlois Foundation

PROGRAM ART OR BE PROGRAMMED
3. http://beandipclothing.com/news/wp-content/uploads/2010/09/andy_warhol_debbie_harry_amiga.jpg

C.E.B. REAS
4. http://reas.com/compendium_a_p/reas_compendium_a_p_0.jpg
5. http://reas.com/compendium_a_p/reas_compendium_a_p_7.jpg 6. need source
6. http://reas.com/compendium_a_p/reas_compendium_a_p_2.jpg
7. http://reas.com/compendium_a_p/reas_compendium_a_p_10.jpg
8. http://reas.com/compendium_a_p/reas_compendium_a_p_14.jpg
9. http://reas.com/compendium_a_p/reas_compendium_a_p_13.jpg

RAFAEL LOZANO-HEMMER
10. http://www.lozano-hemmer.com/image_sets/the_company_of_colours/ny_2009/jpgs/CompanyColours01.jpg
11. http://www.lozano-hemmer.com/image_sets/please_empty_your_pockets/manchester/jpgs/PEYP_PM_26068.jpg

JER THORP
12. http://www.flickr.com/photos/blprnt/3291283918/in/set-72157614008027965/
13. http://www.flickr.com/photos/blprnt/3291287830/in/set-72157614008027965/

GOLAN LEVIN
14. http://www.tmema.org/mis/

MARIUS WATZ
15. http://www.flickr.com/photos/watz/3211511308/in/set-72157612719752007
16. http://www.flickr.com/photos/watz/sets/72157594387603246/

AARON KOBLIN
17. http://www.bicyclebuiltfortwothousand.com/

PATENT FILING DOCUMENTATION, GEODESIC STRUCTURE
18. http://inventors.about.com/od/weirdmuseums/ig/Geodesic-Dome/Geodesic-Dome-Drawing-3.htm

SUBVERT
19. http://www.phreedom.org/cyberpunk-style-guide/shot0741.jpg
20. http://www.flickr.com/photos/ramblingrovers/5249899227/in/photostream/
21. http://upload.wikimedia.org/wikipedia/commons/f/f6/Duchamp_Fountaine.jpg

ROBIN HEWLETT AND BEN KINSLEY
22. http://bkinsley.com/wordpress/wp-content/uploads/2011/08/swav8.jpg
23. http://bkinsley.com/wordpress/wp-content/uploads/2011/08/swav5.jpg
24. http://bkinsley.com/wordpress/wp-content/uploads/2011/08/swav1.jpg
25.http://benkinsley.com/street-with-a-view/

SEBASTIAN BRAJKOVIC
26, 27. http://sebastianbrajkovic.com

JULIUS VON BISMARCK
28–32. http://www.juliusvonbismarck.com/bank/index.php?/projects/fulgurator-action/

HACKER'S MANIFESTO
33. http://www.robdurdle.com/2011/04/11/the-conscience-of-a-hacker-a-hackers-manifesto/

PAUL VANOUSE
34–36. http://www.paulvanouse.com/sic.html

JULIAN OLIVER AND DANJA VASILIEV
37–40. HTTP://NEWSTWEEK.COM/OVERVIEW

MARCO DONNARUMMA
41. http://marcodonnarumma.com/works/golden-shield-music/

THE GREENEYL
42–45. http://www.thegreeneyl.com/jammer-horn

BOREDOM RESEARCH
46–48. http://www.realsnailmail.net/

CITIZEN SCIENCE
49. http://upload.wikimedia.org/wikipedia/en/8/8b/Foldit.png

CESAR HARADA
50. https://sites.google.com/a/opensailing.net/protei/v
51. https://sites.google.com/a/opensailing.net/protei/v/003

HEHE HELEN EVANS AND HEIKO HANSEN
52–54. http://hehe.org3.free.fr

CRITTER
55. http://farm6.static.flickr.com/5246/5380155282_c39322edfa_o.jpg
56. http://farm3.static.flickr.com/2624/3917495479_b8cc04894a_o.jpg
57. http://farm6.static.flickr.com/5125/5380155466_628267b734_o.jpg
58. http://farm3.static.flickr.com/2484/3917496421_b4e9d820bc_o.jpg
59. http://farm3.static.flickr.com/2559/3917495275_ef8212148f_o.jpg
60. http://farm6.static.flickr.com/5208/5379551667_3447c7b908_o.jpg

CENTER FOR POSTNATURAL HISTORY
61. http://postnatural.org/blog/wp-content/uploads/2010/08/NTS_0025_crop.jpg
62. http://postnatural.org/blog/wp-content/uploads/2011/04/CPNH_front.jpg

INSTITUTE FOR FIGURING
63, 64. Courtesy The Institute for Figuring: http://www.theiff.org/

ARTIST IN WHITE COATS
65. Courtesy Philip Ross, CRITTER: http://crittersalon.blogspot.com/
66. http://www.curatedobject.us/the_curated_object_/2008/01/exhibitions-n-2.html
67, 68. Courtesy Philip Ross, CRITTER: http://crittersalon.blogspot.com/

ELEKTRO THE MOTOMAN
69. http://cyberneticzoo.com/?p=1304

BALLENGÉE
70–74. Courtesy Brandon Ballengée: http://www.brandonballengee.com/

ESPARZA
75–78. Courtesy Gilberto Esparza: http://www.plantasnomadas.com/

PHIL ROSS
79, 80. Courtesy Philip Ross: http://www.philross.org/
81. http://billhoss.phpwebhosting.com/ross/typo3temp/abadff07b3.jpg
82. Courtesy Philip Ross: http://www.philross.org/

BCL
83. http://www.common-flowers.org/images/TissueCulture017.jpg
84. http://www.common-flowers.org/images/georg-smelling-blue-carnation.jpg
85. http://www.common-flowers.org/images/BlueCarnation.jpg

KATHY HIGH
86, 87. Courtesy Kathy High: http://vampirestudygroup.com/bloodwars/

CREATURES EVOLVED FOR WALKING
88. http://www.karlsims.com/evolved-virtual-creatures.html

FERNANDO ORELLANA
89. http://fernandoorellana.com/press_ready/Sleep_waking_hor_EMPAC.tif
90. http://fernandoorellana.com/press_ready/daydreaming.jpg
91. http://fernandoorellana.com/press_ready/flying.jpg

INTERVIEW WITH
JOHANNES GRENZFURTHNER
92. http://www.flickr.com/photos/laughingsquid/1501247743/
93. http://www.flickr.com/photos/laughingsquid/2889699191/
94. http://www.flickr.com/photos/laughingsquid/3977147121/
95. http://www.flickr.com/photos/monochrom/5155258997/
96. http://www.flickr.com/photos/weexinsitu/5157412609/
97. http://www.flickr.com/photos/samuel-huron/5158610043/

MATTHEW KENYON & DOUGLAS EASTERLY (SWAMP) + TIAGO RORKE
98, 99. Courtesy SWAMP + Tiago Rorke: http://www.tardigotchi.com/
100. http://www.tardigotchi.com/files/gimgs/6_tardigotchi1.jpg
101. http://www.tardigotchi.com/files/gimgs/6_tardigotchi3.jpg
102. http://www.tardigotchi.com/index.php?/projectdetails/physical/

AGNES MEYER-BRANDIS
103–107. Courtesy Agnes Meyer-Brandis

SYMBIOTICA
108–121. Courtesy SymbioticA: http://www.symbiotica.uwa.edu.au/

MAKER MOMENT
122. http://www.flickr.com/photos/escher_47/134594308/
123. http://www.youtube.com/watch?v=APg1ELLZK-Q

JAMES L. ACORD
124. http://en.wikipedia.org/wiki/File:James_L_Acord_in_front_of_FFTF,_Hanford,_WA,_USA.png

MACHINE PROJECT
http://www.flickr.com/photos/machineproject/
125. Courtesy Machine Project: http://machineproject.com/
126. http://www.flickr.com/photos/machineproject/5489658403/in/set-72157626051602587
127, 128. Courtesy Machine Project: http://machineproject.com/
129. http://www.flickr.com/photos/machineproject/4306138671/in/set-72157623040287084
130. Courtesy Machine Project: http://machineproject.com/

THOMAS THWAITES
131–135. Courtesy Thomas Thwaites: http://www.thomasthwaites.com/

JONAH BRUCKER-COHEN
136–138. Courtesy Jonah Brucker-Cohen and Katherine Moriwaki: http://www.scrapyardchallenge.com/
139. Courtesy STUDIO for Creative Inquiry
140. Courtesy Jonah Brucker-Cohen and Katherine Moriwaki: http://www.scrapyardchallenge.com/

JON COHRS
141, 142. Courtesy Jon Cohrs: http://urban-prospecting.net/

FREE ART AND TECHNOLOGY (FAT), OPENFRAMEWORKS, THE GRAFFITI RESEARCH LAB, AND THE EBELING GROUP
143. Courtesy F.A.T. www.fffff.at
144. http://eyewriter.org/images/data/P1060726.JPG
145. http://eyewriter.org/images/data/RoundOne/P1060130.JPG
146. Courtesy F.A.T. www.fffff.at

GENERAL DYNAMICS 2063 AD TIME CAPSULE
147. http://boingboing.net/2008/04/15/retrofuture-space-vi.html

OVERVIEW EFFECT
148. http://en.wikipedia.org/wiki/File:Neuromancer_%28Book%29.jpg
149. http://en.wikipedia.org/wiki/File:Teilhard_de_Chardin%281%29.jpg
150. http://upload.wikimedia.org/wikipedia/commons/a/a8/NASA-Apollo8-Dec24-Earthrise.jpg

TOMÁS SARACENO
151. Courtesy Tómas Saraceno, Andersen's Contemporary, Tanya Bonakdar Gallery, and pinksummer contemporary art.
152. Courtesy Tómas Saraceno
153. Courtesy Tómas Saraceno and Tanya Bonakdar Gallery

TOMAS SARACENO 2
154. http://arttattler.com/Images/NorthAmerica/Minnesota/WalkerArtCenter/Tomas%20Saraceno/32SW-Stay-Green.jpg

DUNNE & RABY
155. http://www.dunneandraby.co.uk/img/projects/large/4-0.jpg
156. http://www.dunneandraby.co.uk/img/projects/large/6-0.jpg
157. http://www.dunneandraby.co.uk/img/projects/large/7-0.jpg

SASCHA POHFLEPP
158. http://www.pohflepp.com/sites/default/files/images/foreverfuture_2.jpg
159. http://pohflepp.plugimi.com/?q=foreverfuture

BRUCE STERLING
160. 600full-mirrorshades:-an-anthology-of-cyberpunk-(paladin-books)-cover.jpg

ATELIER VAN LIESHOUT
161–163. Courtesy Atelier Van Lieshout: http://www.ateliervanlieshout.com/

ETOY
164. http://www.res-qualia.net/files/641.jpg

INTERMEDIARY - SCIENTIFIC EVANGELIST
165–171. http://commons.wikimedia.org/

PIONEER PLAQUE
172. http://en.wikipedia.org/wiki/File:Pioneer10-plaque.jpg

TIMELINE
1. http://en.wikipedia.org/wiki/File:RothwellMaryShelley.jpg
2. http://en.wikipedia.org/wiki/File:Wheatstone_Charles_drawing_1868.jpg
3. http://www.earlsoham.suffolk.sch.uk/history.html
4. http://www.popsci.com/files/threecovers_reflect.jpg
5. http://www.animated-divots.net/images/phenakistoscope.jpg
6. http://en.wikipedia.org/wiki/File:Popular_Mechanics_Cover_Vol_1_Issue_1_11_January_1902.jpg
7. http://en.wikipedia.org/wiki/File:GBallaArt.jpg
8. http://online.wsj.com/article/SB10001424052748703673604575550201949192336.html
9. http://en.wikipedia.org/wiki/File:Bauhaus.JPG
10. http://en.wikipedia.org/wiki/File:Replica-of-first-transistor.jpg
11. http://www.holtzecompanies.com/images/RADIO%20SHACK%20-%20MORNINGSIDE%201968.jpg
12. http://en.wikipedia.org/wiki/File:L430xH465_jpg_Schaeffer_big-2eb70.jpg
13. http://en.wikipedia.org/wiki/File:GPS_on_smartphone_cycling.JPG
14. http://blog.audiojungle.net/wp-content/uploads/lespaul.jpg
15. http://en.wikipedia.org/wiki/File:Eniac.jpg
16. http://en.wikipedia.org/wiki/File:Polaroid_Highlander_Model_80A_%28John_Kratz%29.jpg
17. http://en.wikipedia.org/wiki/File:Norbert_wiener.jpg
18. http://en.wikipedia.org/wiki/File:Kodak_front_up.jpg
19. http://eamusic.dartmouth.edu/~music3/images/cage-williamsmix.gif
20. http://en.wikipedia.org/wiki/File:Sputnik_1.jpg
21. http://paleo-future.blogspot.com/2007/05/closer-than-we-think-1958-1963.html
22. http://en.wikipedia.org/wiki/File:Xerox_914.jpg
23. http://www.wired.com/thisdayintech/tag/cp-snow/
24. http://en.wikipedia.org/wiki/File:Spacewar!-PDP-1-20070512.jpg
25. http://mysite.pratt.edu/~llaurola/cg550/zajac.jpg
26. http://design.osu.edu/carlson/history/images/ivan-sutherland.jpg
27. http://www.paleofuture.com/blog/2007/7/27/general-dynamics-astronautics-time-capsule-1963.html
28. http://www.atariarchives.org/deli/basic.php

29. http://www.balticmill.com/whatsOn/past/
ExhibitionDetail.php?exhibID=97
30. http://www.tate.org.uk/images/
cms/16510w_howarth_03alt1.jpg
31. http://en.wikipedia.org/wiki/File:STIn
Beauty.jpg
32. http://studioseen.blogspot.com/
33. http://en.wikipedia.org/wiki/
File:Leonardolowres.jpg
34. http://www.justinfarrar.com/wp-content/
uploads/2010/09/whole-earth-catalog.jpg
35. http://laurenfenton.com/?p=1818
36. http://www.spatialagency.net/database/
ant.farm
37. http://en.wikipedia.org/wiki/Moon_Museum
38. http://en.wikipedia.org/wiki/File:The_Earth_
seen_from_Apollo_17.jpg
39. http://www.fondation-langlois.org/html/e/
page.php?NumPage=2155
40. need source
41. http://en.wikipedia.org/wiki/File:Benoit_
Mandelbrot_mg_1804-d.jpg
42. http://en.wikipedia.org/wiki/File:Hp_
laserjet_4200dtns.jpg
43. http://www.atariarchives.org/artist/
sec28.php
44. http://en.wikipedia.org/wiki/File:Apple_II_
IMG_4218-black.jpg
45. http://en.wikipedia.org/wiki/File:Star_Wars_
Logo.svg
46. science.howstuffworks.com/ voyager4.htm
47. http://en.wikipedia.org/wiki/File:Voyager.jpg
48. http://en.wikipedia.org/wiki/File:Xerox_
Star_8010.jpg
49. http://en.wikipedia.org/wiki/File:Blade_Run-
ner_poster.jpg
50. http://en.wikipedia.org/wiki/
File:Macintosh_128k_transparency.png
51. http://en.wikipedia.org/wiki/File:Dennis-
hopper-andy-warhol-at-table-1963.jpg
52. http://en.wikipedia.org/wiki/
File:TimBernersLee-2009.jpg
53. http://en.wikipedia.org/wiki/
File:Photoshopcs5.png
54. http://en.wikipedia.org/wiki/File:Batman_
Digital_Justice_cover.jpg
55. http://en.wikipedia.org/wiki/File:Linux-x86-
under-qemu.png
56. http://en.wikipedia.org/wiki/File:Octopod_
by_syntopia.jpg
57. http://en.wikipedia.org/wiki/File:Adobe_
Flash_CS5.jpg
58. http://en.wikipedia.org/wiki/File:Eyebeam-
2008-kr.jpg
59. http://en.wikipedia.org/wiki/File:Dna-split.png
60. http://en.wikipedia.org/wiki/File:CC-logo.svg
61. http://en.wikipedia.org/wiki/Arduino
62. http://en.wikipedia.org/wiki/
File:GoogleLunarXPRIZE_Logo.png

CONTRIBUTORS

RÉGINE DEBATTY is a blogger, curator and critic whose work focuses on the intersection between art, science and social issues. www.we-make-money-not-art.com

CLAIRE L. EVANS is a writer, science journalist, science-fiction critic, and the author of Universe, a blog addressing the intersections between science and culture. She is also an artist and musician in the band YACHT. www.clairelevans.com

ANDREA GROVER is a curator, artist and writer. She is the founder of Aurora Picture Show, Houston, and has curated exhibitions on art, technology, and collectivity for apexart, New York, and Miller Gallery at Carnegie Mellon University. She is presently Associate Curator at Parrish Art Museum, Southampton, New York. www.andreagrover.com

PABLO R. GARCIA is the founder and principal of POiNT, a collaborative and multidisciplinary research studio based in Pittsburgh. POiNT is dedicated to experiments in the spatial arts—architecture, design, and the visual and performing arts, in a variety of scales from the portable to the urban. www.pointprojects.com

THUMB is a Brooklyn and Baltimore-based graphic design office that was established as a partnership between Jessica Young and Luke Bulman in 2007. Thumb is fond of fluorescent inks, microscopic art, live and immediate processes, color, Ebay, shape, very glossy paper, discs, surprises, diagrams, rainbow paper, and awkward transitions.

POLLY KOCH, copy editor

Support from The Andy Warhol Foundation for the Visual Arts Curatorial Fellowships program.

STUDIO for Creative Inquiry Staff: Golan Levin, Director; Marge Myers, Associate Director; Stephanie Rael, temporary Business Manager, Amisha Gadani, Artist Research Associate; Jonathan Minard, Research Associate.

Miller Gallery Staff: Astria Suparak, Director; Margaret Cox, Graphics and Office Coordinator; Tesar Freeman, Exhibitions Coordinator.

Student Assistants: Aswin Widjaya, Brier Avil, Hanah Lee Ho, Jennifer Kang, Drew Lightfoot, Lara Y. Mann, Lucas Martin, Haydee J. Naula, Gabriela Schmulevich, Regina Son, Audrey Tse, David Yen.

Thanks: Cesar Harada, Jonah Brucker-Cohen, Michael Mandiberg, Patricia Maurides, Michael Pisano, Burr Settles, Natalie Settles, Dan Wilcox, Garth Zeglin, Jon Rubin, Rich Pell, Fred C. Adams, Julian Oliver and Danja Vasiliev, Adam Zaretsky, Jerry Juarez, and Johannes Grenzfurthner.

ACKNOWLEDGMENTS

We extend our deepest gratitude to Astria Suparak, Director of the Miller Gallery at Carnegie Mellon University, Golan Levin, Director of the STUDIO for Creative Inquiry, and Marge Myers, Associate Director. This triumvirate of individuals provided leadership, and created the kind of environment necessary to allow this experimental endeavor to flourish. STUDIO staff, Amisha Gadani, Artist Research Associate, and Jonathan Minard, Research Associate, were our sounding boards, tour guides, and library companions. For welcoming us into their zones during our campus field trips, we thank Lowry Burgess, Professor of Art and Disinguished Fellow at the STUDIO; Richard Pell, Assistant Professor of Art, fellow in the STUDIO for Creative Inquiry and Director of The Center for PostNatural History; and Garth Zeglin, Project Specialist at Robotics Institute. And for providing feedback on the fifth day of our sprint, we are grateful to Burr Settles, Patricia Maurides, Garth Zeglin, Natalie Settles.

Hosts: STUDIO for Creative Inquiry and Miller Gallery at Carnegie Mellon University.

WORD USAGE FREQUENCY IN THIS PUBLICATION

art
work
science
technology
first
artist
computer
use
project
hacker
new
one
design
year
image
process
system
source
research
human
future
cultural
time
public
practice
scientific
hack
blood
engineer
develop
culture
media
through
software
make
method
world
scientist
electronic
robot
current
become
release
see
piece
form
digital
center
create
space

social
artistic
many
way
move
laboratory
rather
technological
cell
call
found
people
invent
caption
data
material
creative
launch
programme
earth
contemporary
http
publish
provide
just
look
behavior
user
term
create
video
program
allow
live
natural
make
understand
ideas
how
physical
open
model
title
limb
lab
change
point
plant
century
better

web
consider
made
internet
life
present
entire
technology
american
view
well
might
network
very
direct
include
position
set
go
approach
exist
dream
relationship
information
product
university
over
device
camera
example
flower
maker
planet
goal
critical
machine
school
common
white
group
workshop
describe
participant
nature
eye
interest
researcher
city
person
fiction

collaboration
operate
tool
found
general
modern
associate
foundation
community
knowledge
sleep
original
wide
experiment
society
write
different
produce
question
diy
establish
perspective
place
context
build
language
study
however
robert
environmental
studio
learn
high
subvert
application
experience
engage
online
back
start
involve
visual
state
invert
practical
activity

COLOPHON

New Art/Science Affinities

Published by Miller Gallery at Carnegie Mellon University + CMU STUDIO for Creative Inquiry
Purnell Center for the Arts
5000 Forbes Avenue
Pittsburgh, PA 15213
www.cmu.edu/millergallery
www.studioforcreativeinquiry.org

Graphic Design
Thumb: Luke Bulman and Jessica Young

New Art/Science Affinities is set in Replica and New Century Schoolbook.

Printed and bound in the United States. All images courtesy the artists unless otherwise noted. All reasonable attempts have been made to identify owners of copyright. Errors or omissions will be corrected in future volumes.

This publication is available in digital form from the publishers.

ISBN: 0977205347

Library of Congress Control Number: 2011931921

First Edition, 2011

www.ingramcontent.com/pod-product-compliance
Ingram Content Group UK Ltd.
Pitfield, Milton Keynes, MK11 3LW, UK
UKHW060016051026
15591UKWH00002B/228
* 9 7 8 0 9 7 7 2 0 5 3 4 9 *